Oxford Engineering Science Texts

Structural mechanics

BY

ANDREW C. PALMER

CLARENDON PRESS · OXFORD
1976

Oxford University Press, Ely House, London W.1

OXFORD LONDON GLASGOW NEW YORK
TORONTO MELBOURNE WELLINGTON CAPE TOWN
IBADAN NAIROBI DAR ES SALAAM LUSAKA ADDIS ABABA
KUALA LUMPUR SINGAPORE JAKARTA HONG KONG TOKYO
DELHI BOMBAY CALCUTTA MADRAS KARACHI

Casebound ISBN 0 19 856127 X
Paperback ISBN 0 19 856128 8

Printed in Great Britain
by J. W. Arrowsmith Ltd., Bristol

Preface

THE idea for this book came from my lectures and tutorials at Cambridge in the years 1967 to 1974. When I wanted to recommend to students a textbook on structural mechanics, I found that there were many books on the subject, but that none of them seemed really satisfactory. Almost always they lacked selectivity, and in page after page covered numerous methods for the solution of problems, including many methods which are already long obsolete (and indeed are nowadays never met with outside textbooks and examination papers), and other methods which are rapidly becoming obsolete. Universities ask more and more of students, and constantly (and of course quite properly) review and update syllabuses and introduce new subjects. It seemed to me that a textbook should be set out to simplify as much as possible, to be as selective as possible, and not to be afraid to leave out ideas that are now no longer important.

A second motivation was the lack of a book which based its approach on the extraordinarily useful concept of virtual work. Most books mention virtual work in passing, but hardly any exploit its full power. One of the things I attempt to do in the present book is to show how powerful a concept it is, and that it can replace a whole series of different special methods for different kinds of problems. A few people still argue that virtual work is somehow 'impractical' and 'academic'. I reject this view: on the contrary, it seems to me that the simplest ideas will be the ones most useful in engineering practice. I came across an illustration of this point about a month before writing this preface. It arose in a construction problem, that of finding the tensions in two cables, pulled by winches, which lift a floating stern ramp into attachment points on a pipe-laying barge. Analysis of several of the different cases of interest had taken several days' work by two engineers using traditional methods, but it was then found that virtual work gave a solution in much less time (about a fifth as much), and that it in fact gave a clearer idea of what were the main factors determining the cable tensions.

I set out to write a short book, which would outline the central ideas of structural mechanics, and would use virtual work as a unifying concept. The book is set at the level of a first- or second-year degree course in structural engineering. Central ideas in mechanics, such as the conditions for a solid body to remain at rest in equilibrium, are explained and discussed afresh, but it is assumed that almost every reader will have met them before. Some use is made of geometry and calculus, but the level of knowledge assumed is again not beyond that of a student starting a university or polytechnic course in engineering. A note on page xii summarizes the mathematical background that different chapters assume.

Solid mechanics, or what used to be called 'strength of materials', is outside the scope of the book, and a knowledge of it is not assumed. A course in solid mechanics is of course an essential part of an engineer's education, but it can come either before or after a course in structural mechanics. In beam theory, for instance, this book first introduces the statical ideas used to describe forces within beams (bending moment, shear force, and so on), then explores geometric ideas such as curvature, and only then examines the consequences of the interaction between statics and geometry. In this last analysis use is made of relationships between curvature and bending moment, and of simple idealizations and descriptions of these relations, but the book does not cover the details of how they in turn relate to the material and cross-section of the beam. A reader who has studied solid mechanics will know what this relationship is; a reader who has not done so merely has to accept that a relationship between curvature and bending moment does exist, and that it can be discovered somehow, whether by theoretical analysis, or by experiment, or by a combination of the two.

The reader already familiar with the subject will be surprised at some of the omissions. They are deliberate, and occur for the reasons explained in the first paragraph of this preface. I have not made any use of the energy methods associated with the names of Engesser and Castigliano, since I feel that the problems for which they are useful are better solved by virtual work. I was more reluctant to leave out moment distribution, but in the end decided to do so. It is an attractive technique, and engineers enjoy using it (which is not a negligible advantage), but its usefulness has been overtaken by the rapid development of computer methods. At least in the developed countries, an engineer who wants to analyse a structure of even moderate complexity nowadays almost always has access to a library of computer programs which will do it for him. Pocket calculators already have programs of this kind. An engineer has to understand what physical ideas such a program is using, and he ought to be able to solve simple problems without its help, but there is much less need than there used to be for facility in the use of methods like moment distribution.

I owe a debt of gratitude to very many colleagues and students, at Cambridge, Liverpool, and Brown Universities. I should especially like to thank John Baker, Chris Calladine, Dan Drucker, Jack Ells, Jacques Heyman, Turan Onat, and Andrew Schofield, for the privilege of discussing mechanics with them. The mistakes in this book are of course my own. I should also like to thank Fred Leckie, the editor of this series, the staff of Oxford University Press, and several anonymous readers for their helpful criticism and encouragement. Finally, I should like to record my gratitude to my wife Jane and my daughter Emily, for their patience and understanding while I have been writing.

ANDREW C. PALMER

Rijswijk, Netherlands
October, 1975

Contents

NOTE ON THE MATHEMATICAL BACKGROUND ASSUMED OF THE READER

A knowledge of elementary calculus (principally differentiation and integration) is essential, together with simple analytic geometry and trigonometry. Additional concepts used in particular chapters are listed below.

Chapter 2: vector notation; vector product (cross product);

Chapter 4, section 4.3: unit vectors used to describe direction;

Chapter 7, sections 7.4 and 7.5: solution of linear differential equations with constant coefficients;

Chapter 7, section 7.5: Fourier series.

1. Introduction

AN engineer concerned with structures finds himself confronted with questions about their mechanical behaviour. The questions might be: how much load can be applied to a certain bridge before it collapses? If a 50 tonne load is applied at one point, will the deflection 2 m away be more than 10 mm? Is there a risk that the framework will buckle? How should it be redesigned to carry a greater load, or to cover a wider span? This book sets out to explain and develop methods which an engineer can apply to problems of this kind. The central ideas come from mechanics, and are supplemented by concepts from geometry and applied mathematics. Although the methods are stated and developed in the context of civil engineering structures, the underlying ideas are in fact common to solid mechanics as a whole, and have applications in many other disciplines, in mechanical and aerospace engineering, in soil mechanics, in geophysics and glaciology, and in biology and medicine.

Figure 1.1 shows a specialized and in some ways unusual structure, the stern ramp of a large semi-submersible barge which lays pipelines in deep water. The pipeline is welded together on the barge, and is supported by the ramp as it leaves the barge and arcs into the sea. The ramp is 100 m long, and has to resist loads of the order of 50 tonnes applied by the pipe, as well as wind and wave loading and dynamic loads induced by the motion of the barge. An engineer who sets out to find what is happening to a structure like this neither wants nor needs to include in his analysis every possible piece of information which would come into a complete description. For instance, whether the structure is painted black or white will almost certainly have no effect on how it responds to loads. A decision to ignore the colour of the paint is based on the engineer's knowledge and experience of mechanics and of physics generally, and he will keep somewhere in the back of his mind the fact that if the structure is in the sun, and is unevenly heated so that one side expands more than the other, then the colour may after all turn out to be highly relevant.

Instead, then, of trying to include everything in his analysis, the engineer works with a *model* or idealization in his mind, a model which includes those aspects which are significant and leaves out those which are not. How one decides this is of course vitally important, and is the most central part of the art and science of structural engineering. There is no one choice for any particular structure, and in every case it depends on the context. If the stern ramp in Figure 1.1 were lifted by a single crane, and one were analysing the lifting system, it might well be adequate simply to idealize the ramp as a *concentrated mass.* If it were lifted by two cranes, then its rotational inertia

FIG. 1.1. Stern ramp and pipe-laying barge.

might be important, and it would be better treated as an extended *rigid body*. Each of these idealizations is useful in a particular context, but neither tells us anything about the forces within the ramp. To learn about these forces, one might idealize the ramp as a *beam* (lumping together the forces across a section of the ramp, and working with their resultant), or go further and idealize it as a *framework*, made up of simple bars hinged together, each bar either in tension or compression, or go further still and take account of the fact that the bars are not freely hinged, and so on. There can never be a single answer, but each is right in its own context, just as would be an accountant's idealization of the ramp as a certain investment generating a certain cash flow in return, or a poet's view of the stern ramp as, say, a heron's beak poised over a lake.

One cannot make a rational choice of idealization without knowing what each choice implies, what each includes and leaves out, and how mechanics is applied to the analysis of each idealized structure. It is this subject that the present book is concerned with.

There is another restriction on the field covered by the book, a restriction prompted by the desire for simplicity, and for mechanical and geometrical principles not to be obscured by mathematical detail. The study is restricted to skeletal structures, which can be represented by a network of elements, each of which can in turn be represented by a line with certain properties. Thus, we might represent one of the tubular elements of the stern ramp by a straight line in a diagram, and assert that the corresponding structural element can carry a certain axial force, but leave out of account any detailed analysis of the distribution of force and stress within the element. This too is an idealization, but a remarkably powerful one, and it turns out that this is nearly always the best way to analyse any complex 'frame-like' structure. One first treats the elements of the frame as flexible lines, and determines the resultant force and moment transmitted by the lines, and then separately analyses the effects of these forces and moments. This idealization is unsuitable for structures which are not frame-like: it would serve very well for a beam or a crane jib, or for the stern ramp, but not for a shell roof or for the hull of the barge in the background of Figure 1.1, which have to be treated by a theory of flexible surfaces, nor for a three-dimensional solid body such as a crankshaft or a butt weld or an embankment. The theory of two-dimensional flexible surfaces is shell theory, and has an extensive theoretical development of its own, but one which applies few ideas beyond the ones you will meet in a comparatively simple context in this book. Three-dimensional bodies are the subject of solid mechanics, and it is solid mechanics too that gives us a theory of individual elements of skeletal structures. Solid mechanics in this sense is outside the scope of the present book, and serves as an input to structural mechanics. It tells us how the amount a metal bar stretches under load depends on how large the load is, on what shape the bar cross-section is, on what material it is made from, and so on. Structural mechanics in turn uses the results of solid mechanics to analyse the behaviour of complete structures. Solid mechanics is described in a number of texts, at different levels: for an introduction, see Crandall and Dahl (1959) or Drucker (1967).

The most important concepts we take from mechanics are those of statics and equilibrium, and it is them that we consider first.

2. Statics

2.1. Introduction, terminology, and notation

IMAGINE a structure which is at rest and is carrying some loads. The forces acting on it together have to satisfy some conditions, for otherwise it cannot remain in equilibrium. For example, a structure acted on by only the following forces cannot be in equilibrium:

force 1: 300 N, acting upwards,
force 2: 700 N, acting downwards,
force 3: 600 N, acting upwards,

since an unbalanced upward force of 200 N would accelerate it. These three forces fail to obey one of the statical conditions that have to be obeyed if the structure is to be in equilibrium. We begin by reviewing these conditions. The concepts of statics are the most important and frequently used concepts of structural mechanics. You will almost certainly have met them before: if not, you ought before going further to go back to a textbook on elementary statics. Vector notation will be used fairly extensively: this too you should have met before.

First, some notation is needed, so that we have a common terminology and so that forces and the points at which they act can be described mathematically. Either a vector notation can be used, or vector quantities like forces can be described in terms of their components with respect to an orthogonal coordinate system. In vector notation, we number the various forces in some convenient way, and call them $\mathbf{P}_1$, $\mathbf{P}_2$, $\mathbf{P}_3$, and so on, so that the ith force is $\mathbf{P}_i$. $\mathbf{P}_1$ is written in bold-face type to indicate that it denotes a vector; in hand-written work it is convenient to underline vector quantities. The points at which forces act are located by position vectors $\mathbf{r}$, measured from some convenient reference point, so that $\mathbf{P}_1$ acts at a point located by $\mathbf{r}_1$, $\mathbf{P}_2$ at a point located by $\mathbf{r}_2$, and so on. In component notation we shall use a right-handed coordinate system, whose axes are labelled 1, 2, and 3 (Figure 2.1). This is nothing more than the familiar x, y, z coordinate system, but it saves trouble later to identify axes by numbers rather than letters. Note that the numbering of the axes is not completely arbitrary: if you hold your right hand with the thumb and middle finger at right angles, and the fore-finger pointing outwards, *away* from the palm of the hand, and at right angles to the other two, then the thumb, fore-finger, and middle finger point in the positive 1-, 2-, and 3-directions of a right-handed system. Each force has three components, one in each of the coordinate directions. The first force, $\mathbf{P}_1$ in vector notation, has components P_{11}, P_{12}, and P_{13} in the 1-, 2-, and 3-directions; the components themselves are scalar quantities, which is

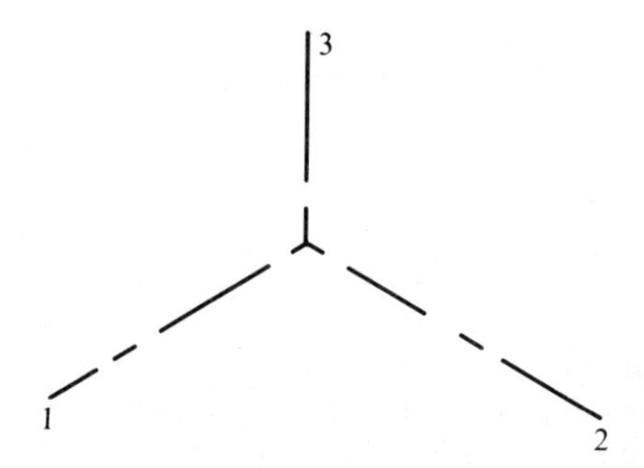

FIG. 2.1. Right-handed system of coordinate axes.

why they are written in italic rather than bold face, and it is the first subscript which indicates which force is being described, and the second which indicates which component it is. Thus, P_{13} is the 3-component (equivalently, z-component) of the first force $\mathbf{P}_1$, and P_{31} is the 1-component (equivalently, x-component) of the third force $\mathbf{P}_3$. Figure 2.2 shows a force $\mathbf{P}_i$ and its three components P_{i1}, P_{i2}, and P_{i3}. Similarly, the components of position vector $\mathbf{r}_i$ are written r_{i1}, r_{i2}, and r_{i3}.

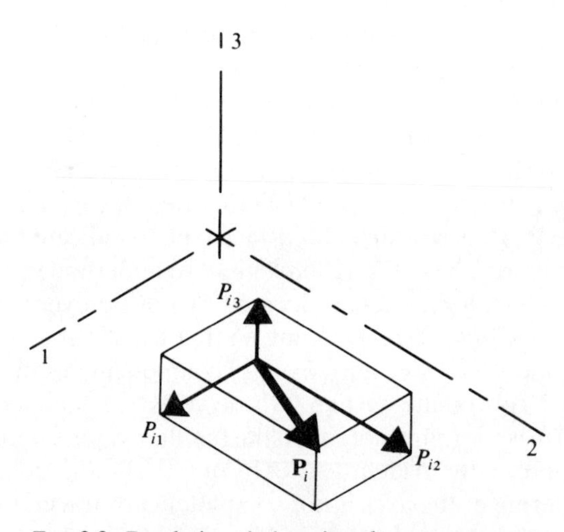

FIG. 2.2. Resolution of a force into three components.

We also need to define the moment of a force about a point. If an arbitrary point A is located by a vector $\mathbf{a}_1$ (components a_{11}, a_{12}, and a_{13}) and a force $\mathbf{P}_1$ acts at a point located by $\mathbf{r}_1$, then the moment of $\mathbf{P}_1$ about A is the vector

$$(\mathbf{r}_1 - \mathbf{a}_1) \times \mathbf{P}_1.$$

Here $\times$ denotes the vector product, sometimes called the cross product to distinguish it from the dot product (or scalar product). In order to interpret this in terms of components, think of the force $\mathbf{P}_1$ resolved into its three components P_{11}, P_{12}, and P_{13}, and the relative position vector $\mathbf{r}_1 - \mathbf{a}_1$ resolved into its three components $r_{11} - a_{11}$, $r_{12} - a_{12}$, and $r_{13} - a_{13}$. Figure 2.3 shows a view along the 1-axis. A component of the moment is counted positive if the moment looking along the corresponding axis acts clockwise. The moment of the force is the sum of the moments of its components. In Figure 2.3, the lever arm of component P_{13} about A is AD, the perpendicular distance from A to the line of action of P_{13}. The magnitude of AD is

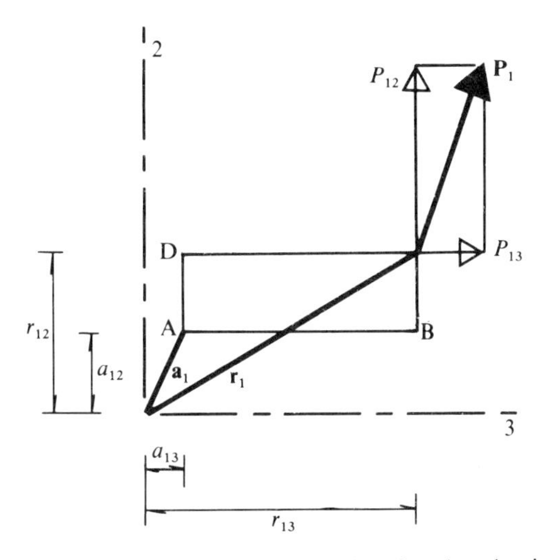

FIG. 2.3. Moment of a force about a point: view along 1-axis.

$r_{12} - a_{12}$ and the corresponding contribution to the 3-component of the moment is $P_{13}(r_{12} - a_{12})$. The lever arm of component P_{12} about A is AB, whose magnitude is $r_{13} - a_{13}$; its contribution to the 3-component of the moment is $-P_{12}(r_{13} - a_{13})$, the minus sign being required because this contribution is counter-clockwise. The resultant component of the moment is

$$(r_{12} - a_{12})P_{13} - (r_{13} - a_{13})P_{12}.$$

If up till now you have only met moments in the context of forces in a single plane, you will recognize this as the moment of $\mathbf{P}_1$ about A in such a system. In general, however, this is only one of three components of the moment of $\mathbf{P}_1$ about A. Figure 2.3 is a view along the 1-axis, and the above calculation gives just the 1-component. It is important not to forget that there are other components. If instead we view the system along the 2-axis (Figure 2.4), and

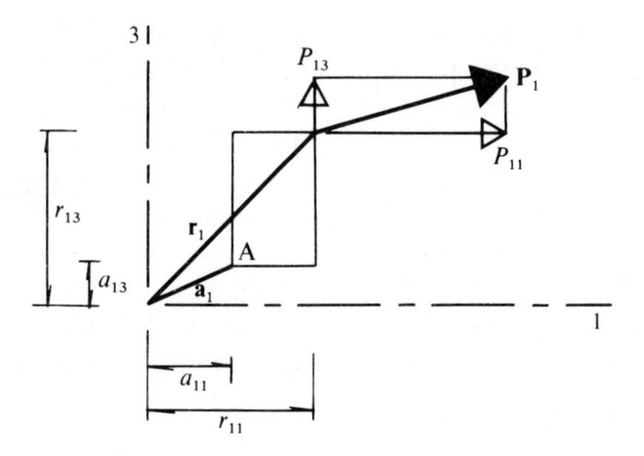

FIG. 2.4. Moment of a force about a point: view along 2-axis.

again take moments for the components, we get the 2-component of the moment, which is

$$(r_{13}-a_{13})P_{11}-(r_{11}-a_{11})P_{13}.$$

The 3-component can be derived in the same way, and is

$$(r_{11}-a_{11})P_{12}-(r_{12}-a_{12})P_{11}.$$

Having derived the first of these components, it is in fact easy to write down the other components. The first subscript is always 1, because all this refers to force 1; the second subscripts permute cyclically 1 2 3 1 2 3 and so on. For example, the first term in the 1-component is $(r_{12}-a_{12})P_{13}$; to get the corresponding term in the 2-component, change 1 to 2, 2 to 3, and 3 to 1 whenever they appear as second subscripts. To get the corresponding term in the third component, repeat the process. The other terms are obtained similarly. (Check this for yourself.)

2.2. Equilibrium

This terminology and notation make it possible to describe the conditions that must be obeyed if a structure is to be in equilibrium. These conditions are:

I. the vector sum of the forces acting on the structure must be zero.
II. the vector sum of the moments about any arbitrary point of the forces acting on the structure, together with any externally applied moments, must be zero.

if the forces are again $\mathbf{P}_1, \mathbf{P}_2, \ldots$, acting at points located by $\mathbf{r}_1, \mathbf{r}_2, \ldots$, and the external moments are $\mathbf{M}_1, \mathbf{M}_2, \ldots$, the conditions are, in vector notation

I. $$\sum_{\text{all forces } i} \mathbf{P}_i = \mathbf{0}, \qquad (2.2.1)$$

$$\text{II.} \qquad \sum_{\text{all forces } i} \{(\mathbf{r}_i - \mathbf{a}) \times \mathbf{P}_i\} + \sum_{\text{all moments } i} \mathbf{M}_i = \mathbf{0}, \qquad (2.2.2)$$

taking moments about an arbitrary point located by **a**.

In component notation, each of these single conditions on a vector sum becomes three conditions written in scalar components. Thus, the 1-component of the vector sum of the forces is $P_{11} + P_{21} + P_{31} + \ldots$, and must be zero. Similarly the 2-component and the 3-component of the vector sum must be zero. Condition I becomes

$$\text{I.} \quad \begin{cases} \sum_{\text{all forces } i} P_{i1} = 0 \\ \sum_{\text{all forces } i} P_{i2} = 0 \\ \sum_{\text{all forces } i} P_{i3} = 0, \end{cases} \qquad (2.2.3)$$

and condition II becomes

$$\text{II.} \quad \begin{cases} \sum_{\text{all forces } i} \{P_{i3}(r_{i2} - a_2) - P_{i2}(r_{i3} - a_3)\} + \sum_{\text{all moments } i} M_{i1} = 0 \\ \sum_{\text{all forces } i} \{P_{i1}(r_{i3} - a_3) - P_{i3}(r_{i1} - a_1)\} + \sum_{\text{all moments } i} M_{i2} = 0 \\ \sum_{\text{all forces } i} \{P_{i2}(r_{i1} - a_1) - P_{i1}(r_{i2} - a_2)\} + \sum_{\text{all moments } i} M_{i3} = 0. \end{cases} \qquad (2.2.4)$$

If condition I is satisfied, the choice of the point to take moments about for condition II is immaterial. If we take moments about an alternative point **b**, then

$$\begin{cases} \sum_{\text{all forces } i} (\mathbf{r}_i - \mathbf{b}) \times \mathbf{P}_i = \sum_{\text{all forces } i} (\mathbf{r}_i - \mathbf{a} + \mathbf{a} - \mathbf{b}) \times \mathbf{P}_i \\ \qquad = \sum_{\text{all forces } i} (\mathbf{r}_i - \mathbf{a}) \times \mathbf{P}_i + (\mathbf{a} - \mathbf{b}) \times \sum_{\text{all forces } i} \mathbf{P}_i \\ \qquad = \sum_{\text{all forces } i} (\mathbf{r}_i - \mathbf{a}) \times \mathbf{P}_i, \end{cases} \qquad (2.2.5)$$

using condition I.

In other words, if condition I holds, the vector sum of the moments of the forces has the same value whatever point the moments are taken about.

The geometrical interpretation of vector addition can be used to express these conditions. Condition I implies that the vectors representing the forces acting on a structure in equilibrium together form a closed polygon, and condition II that the vectors representing the corresponding moments also form a closed polygon. Neither polygon is necessarily plane. In Figure 2.5(a), for instance, four forces together keep in equilibrium a section of pipe suspended in water: resultant forces $\mathbf{P}_1$ and $\mathbf{P}_2$ across the ends of the section, its weight $\mathbf{P}_3$, and a resultant force $\mathbf{P}_4$ exerted by the surrounding water on the outer surface of the pipe. Condition I tells us that vectors representing the four forces must together form a closed polygon like the one illustrated in Figure 2.5(b).

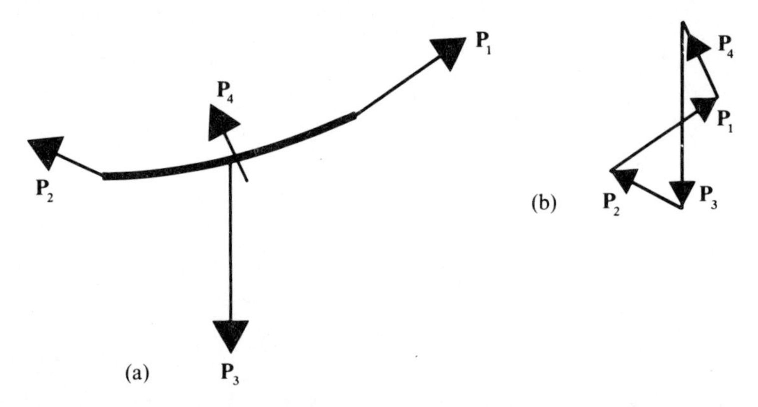

FIG. 2.5. (a) Length of pipe in equilibrium under the action of four forces. (b) Force polygon.

Various special results can be derived. The only ones of sufficient generality to be worth remembering are the following. First, if a body is in equilibrium under only two forces, they must be equal in magnitude, opposite in direction, and their lines of action must coincide. Secondly, if a body is in equilibrium under the action of only three forces, they must all be parallel to a single plane, because the three sides of a triangle lie in the same plane. In addition the lines of action of the three forces must either be parallel or must all pass through the same point.

2.3. External forces on structures

Consider first of all the forces between a structure and whatever foundation supports it, without for the moment asking what forces are set up within the structure itself. The forces must obey the conditions stated in the last section for the equilibrium of the whole structure. To start with, consider two plane structures in which all the forces and all their points of application lie in the same plane.

Example 2.3.1. A bridge truss, illustrated in Figure 2.6, rests on two supports. It carries a 100 kN vertical load at the quarter-span. The left-hand support

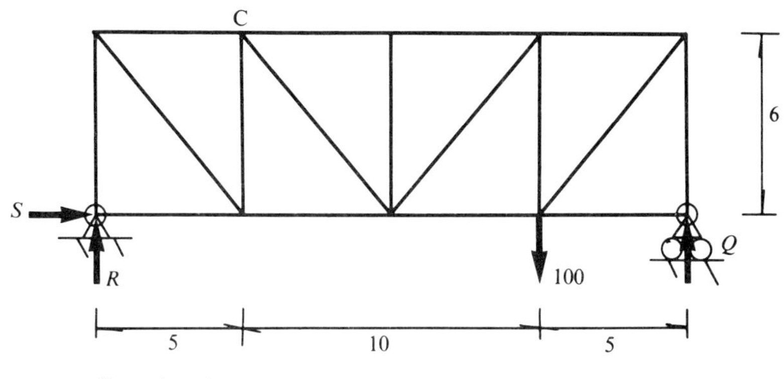

FIG. 2.6. Bridge truss.

can transmit both horizontal and vertical reactions between the truss and its foundation, but the right-hand support can transmit only vertical reactions. Neither support can transmit moments. What are the support reactions?

There are three unknown forces that act on the structure:
Q, the vertical reaction at the right-hand support,
R, the vertical reaction at the left-hand support, and
S, the horizontal reaction at the left-hand support.

These are the forces that the foundation exerts on the structure; the forces that the structure exerts on the foundation are equal and opposite. Condition I tells us that the vector sum of the forces on the structure must be zero. Since its vertical component is zero

$$Q+R-100=0, \tag{2.3.1}$$

and since its horizontal component is zero

$$S=0. \tag{2.3.2}$$

The condition that the resultant force in the third direction, normal to the plane of Figure 2.6, be zero, is automatically satisfied, and tells us nothing new. So far we have enough information to determine S, but not Q or R. Now take moments about the left-hand support. Equilibrium requires that

$$(-Q)(20)+(100)(15)+(R)(0)+(S)(0)=0, \tag{2.3.3}$$

and therefore

$$Q=75 \text{ kN},$$

and, substituting into (2.3.1),

$$R=25 \text{ kN}.$$

In terms of components of a moment vector, the left-hand side of eqn (2.3.3) is the component normal to the plane of Figure 2.6 of the vector sum of the moments, and the other components are identically zero. It was particularly convenient to take moments about the left-hand support, because only one of the unknown forces has any moment about that point, and so only one unknown enters the equation, and can be found immediately. The solution is of course exactly the same whatever point we take moments about. If instead we take moments about point C in Figure 2.6, we have

$$(-Q)(15)+(100)(10)+(R)(5)+(-S)(6)=0. \tag{2.3.4}$$

Using eqn (2.3.2), the equilibrium equations (2.3.1) and (2.3.4) are

$$Q+\ R=\ 100 \tag{2.3.5}$$

and

$$15Q-5R=1000,$$

which can be solved as simultaneous equations to give the same results as before.

In this instance, the unknown reactions between the structure and its foundation can be found from the equilibrium conditions alone. We say that the reactions are *statically determinate.* Later we shall see that this is not always so.

Whether or not the reactions on a particular structure are statically determinate usually depends on the support conditions, but not on the details of the loading. Consider again the structure studied in the previous example, but now let the loading be different (Figure 2.7). Again apply the

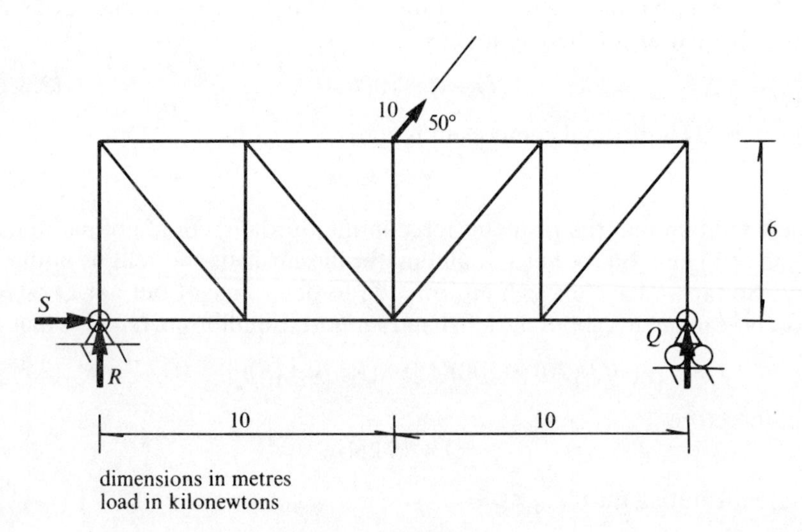

FIG. 2.7. Bridge truss.

equilibrium conditions. Since the vertical component of the vector sum of the forces is zero

$$Q + R + 10 \sin 50° = 0, \tag{2.3.6}$$

and since the horizontal component is zero

$$S + 10 \cos 50° = 0, \tag{2.3.7}$$

and taking moments about the left-hand support

$$(-Q)(20) - (10 \sin 50°)(10) + (10 \cos 50°)(6) + (R)(0) + (S)(0) = 0, \tag{2.3.8}$$

and so

$$Q = 3 \cos 50° - 5 \sin 50° \text{ kN}$$

$$R = -3 \cos 50° - 5 \sin 50° \text{ kN}$$

$$S = -10 \cos 50° \text{ kN}.$$

Finally, if the loadings of Figure 2.6 and Figure 2.7 act together, as in Figure 2.8, the same method of solution gives

$$Q = 75 + 3 \cos 50° - 5 \sin 50° \text{ kN}$$

$$R = 25 - 3 \cos 50° - 5 \sin 50° \text{ kN}$$

$$S = -10 \cos 50° \text{ kN}.$$

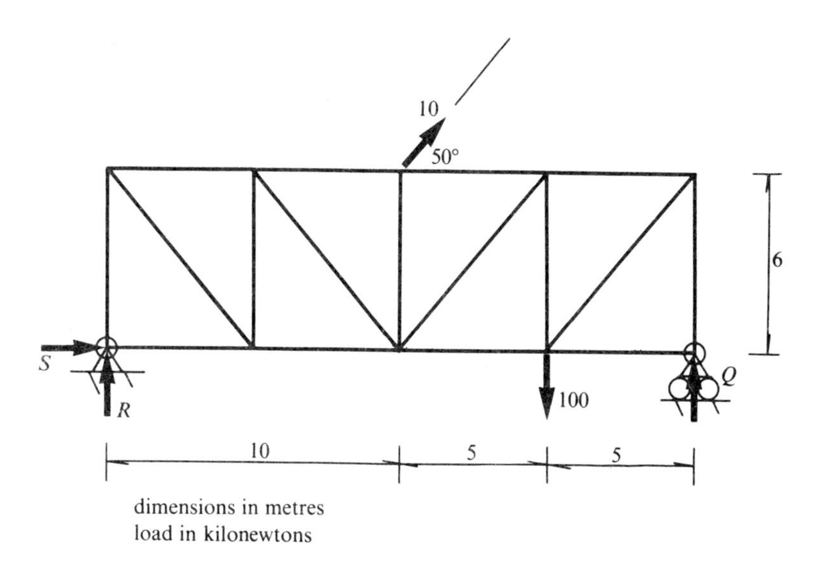

FIG. 2.8. Bridge truss.

Comparison shows that each support reaction induced by the combined loading is the sum of the corresponding individual values for the loadings of Figures 2.6 and 2.7 acting separately. This is an example of what is called the principle of superposition, and in this form it applies generally to statically determinate structures. Chapter 8 discusses it further.

Example 2.3.2. A cantilevered beam of length L is built-in at the left-hand end and free at the other end (Figure 2.9). It carries a distributed vertical load whose intensity varies along its length, so that at a distance x from the left-hand end the load intensity is w(x) per unit length. What are the reactions at the left-hand end?

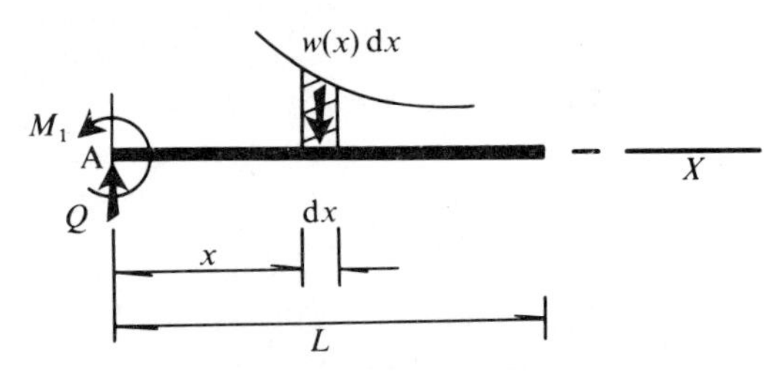

FIG. 2.9. Cantilevered beam.

So that the beam shall be in equilibrium, we expect at the left-hand end a reaction force with vertical and horizontal components. By itself, however, a force at the end A cannot keep the beam in equilibrium. The distributed loading will have a moment about the left-hand end, a moment which the end reaction can do nothing to balance, because, no matter what value it has, its moment about A is zero. It follows that there has to be a moment reaction at A, which will take part in the moment equilibrium condition II; denote this moment M_1. Another way of looking at it is to see that if the built-in end at A were replaced by a hinge, the beam could not be in equilibrium, and would instead simply drop down by rotating about A.

Concentrate attention first on an infinitesimal element dx of the beam, at a distance x from A. The load on the element is $w(x)\,dx$, the load intensity $w(x)$ multiplied by the element length dx. The moment of this force about A is $xw(x)\,dx$, the lever arm x from A to the line of action of the force, multiplied by its magnitude $w(x)\,dx$. These are the force and moment components corresponding to the element dx. If we consider the whole beam, the corresponding force and moment quantities are got by summing the contributions from infinitesimal elements, mathematically by integrating from $x=0$ at A to $x=L$. Applying the equilibrium conditions, since the resultant vertical force on the whole beam is zero,

$$Q=\int_0^L w(x)\,dx, \tag{2.3.9}$$

and since the resultant horizontal force is zero,

$$R = 0, \tag{2.3.10}$$

and since the resultant moment about A is zero,

$$M_1 = \int_0^L xw(x)\,\mathrm{d}x. \tag{2.3.11}$$

Suppose, for instance, that the cantilever is 5 m long, that over the first 2 m ($x < 2$) the load intensity is zero, and that over the outer 3 m ($2 < x < 5$) the load intensity increases linearly from 10 kN/m at $x = 2$ to 30 kN/m at $x = 5$. Then

$$w(x) = \begin{cases} 0 & x < 2 \\ -10 + 10x & x > 2, \end{cases} \tag{2.3.12}$$

and so

$$Q = \int_2^5 (-10 + 10x)\,\mathrm{d}x = 75 \text{ kN}$$

$$M_1 = \int_2^5 (-10x + 10x^2)\,\mathrm{d}x = 285 \text{ kN m}.$$

Example 2.3.3. A bridge truss, illustrated in Figure 2.10, rests on three supports. It carries a 100 kN load at the quarter-span. The left-hand support can transmist both horizontal and vertical reactions, but the other two supports can only transmit vertical reactions. What are the support reactions?

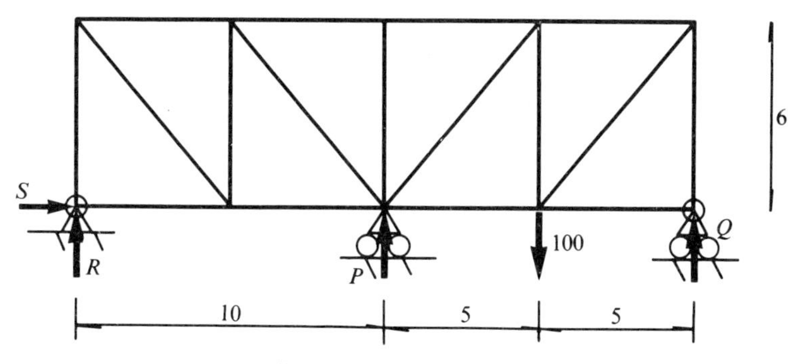

FIG. 2.10. Bridge truss.

This is the same truss as in Example 2.3.1, but now it has three supports instead of two.

Call the reactions P, Q, R, and S, as indicated. Since there is no other horizontal force on the structure besides S, equilibrium requires that

$$S=0. \tag{2.3.13}$$

Vertical equilibrium requires that

$$P+Q+R=100. \tag{2.3.14}$$

Taking moments about the left-hand support

$$10P+20Q=1500. \tag{2.3.15}$$

At this point there are two equations relating three unknown quantities P, Q, and R. There are not yet enough equations to solve for the unknowns. The natural thing to do is to find a third equation by taking moments about some other point. There will then be three equations for three unknowns, and it ought to be possible to solve for the three unknowns. Suppose, then, that we take moments about the right-hand support, which gives us

$$10P+20R=500. \tag{2.3.16}$$

However, if we try to solve eqns (2.3.14), (2.3.15), and (2.3.16), following the usual method for simultaneous equations, difficulties appear. Eliminate R between eqns (2.3.16) and (2.3.14), by dividing eqn (2.3.16) by 10 and subtracting the result from eqn (2.3.14) multiplied by 2, which gives

$$P+2Q=150. \tag{2.3.17}$$

If we now attempt to eliminate either P or Q between eqns (2.3.17) and (2.3.15), we find it to be impossible, because eqn (2.3.17) is nothing more than eqn (2.3.15) divided by 10, and gives us no additional information. The reason for this was indicated in general terms in Section 2.2; if condition I is satisfied, and the resultant moment about one point (say the left-hand support) is zero, then the resultant moment about any other point is automatically zero. Indeed, eqn (2.3.16) can be got from eqns (2.3.14) and (2.3.15) without using mechanics at all, simply by multiplying eqn (2.3.14) by 20 and subtracting eqn (2.3.15) from the result.

It seems, then, that in this instance we cannot determine P, Q, and R, at least not by the equilibrium conditions alone. This conclusion is correct. Figure 2.10 shows an example of what is called a statically indeterminate structure, one for which statics does not by itself tell us what the forces on the structure are.

The physical reason for the indeterminacy of the structure is that it has one more support than it needs if it is to be in equilibrium. If it only had one support, say the left-hand one, then it could not be in equilibrium at all: the load would make it rotate about the support, and it would behave as a mechanism. If it had two supports, as in Example 2.3.1, it would be just in equilibrium, and the support reactions could be found from the equilibrium conditions. A third support is 'unnecessary', in the sense that the structure can be in equilibrium without it. Structural mechanics calls such a support *redundant.*

Even though there are only two equilibrium equations in Example 2.3.3, and they are not enough to make it possible to find P, Q, and R, they must nevertheless of course be obeyed if the structure is to be in equilibrium, and do therefore give us useful information. In particular, if we had some independent way of finding one of the reactions, say Q, these two equations would enable us to find the remaining two reactions. Because knowledge of one of the reactions is enough for us to be able to find the others by statics, we call such a structure once redundant.

These three examples have been concerned with plane structures loaded by coplanar forces. You ought now to be able to solve Problems 2.1–2.5 (page 20). The following examples are concerned with structures loaded by more general systems of forces.

Example 2.3.4. Figure 2.11(a) is a plan view of a three-legged coffee table, whose legs A, B, and C rest on a horizontal floor. Each leg is on a wheeled castor, and so the reaction exerted by the floor on each leg can only be vertical. An object weighing 150 N rests on the table at D. What are the reactions between the three legs and the floor?

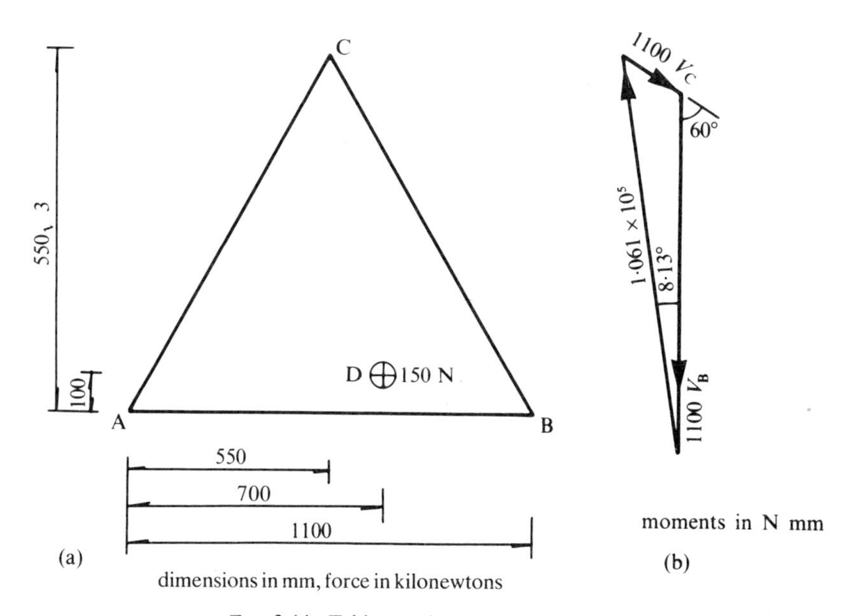

FIG. 2.11. Table carrying concentrated load.

Call the vertical reactions V_A, V_B, and V_C. From vertical equilibrium

$$V_A + V_B + V_C = 150. \qquad (2.3.18)$$

Take moments about A. Reaction V_A passes through A and therefore has no moment about it. Reaction V_B is vertical, and its lever arm AB is horizontal; its moment about A is perpendicular both to V_B and to the lever arm, and is therefore horizontal and at right angles to AB. Its magnitude is 1100 V_B. Similarly, reaction V_C is vertical: the corresponding moment vector, of magnitude 1100 V_C, is horizontal and at right angles to AC. Finally, the weight of 150 N acts vertically at D. Its moment about A is also horizontal, and at right angles to AD, which makes an angle $\arctan(100/700) = 8{\cdot}13°$ with AB. The length of AD is $\sqrt{(700^2 + 100^2)} = 707{\cdot}1$ mm, and so the magnitude of the moment about A of the 150 N weight is $(707{\cdot}1)(150) = 1{\cdot}061 \times 10^5$ N mm.

Equilibrium requires that the vector sum of the three non-zero moments about A be zero. The three individual moment vectors form the closed triangle shown in Figure 2.11(b). In this triangle we know one side and all three angles. It is then simple to determine the two unknown sides, by trigonometry or graphically, and then

$$1100\ V_B = 9{\cdot}64 \times 10^4 \text{ N mm}$$

and so

$$V_B = 88 \text{ N},$$

and

$$1100\ V_C = 1{\cdot}73 \times 10^4 \text{ N mm}$$

and so

$$V_C = 16 \text{ N},$$

and finally, using eqn (2.3.18),

$$V_A = 46 \text{ N}.$$

You can instead derive the reactions by taking moments about a different support. What happens if you take moments about a point not at one of the supports, but on a line joining two of them?

Example 2.3.5. Figure 2.12 shows a section ABCD of pipe connecting two pressure vessels in a chemical plant. It consists of three straight segments, joining points A(0, 0, 0), B(0, 2, 0), C(2, 2, 0), and D(5·6, 6·8, 2·5), which are located by an orthogonal coordinate system whose 3-axis is vertical. At A

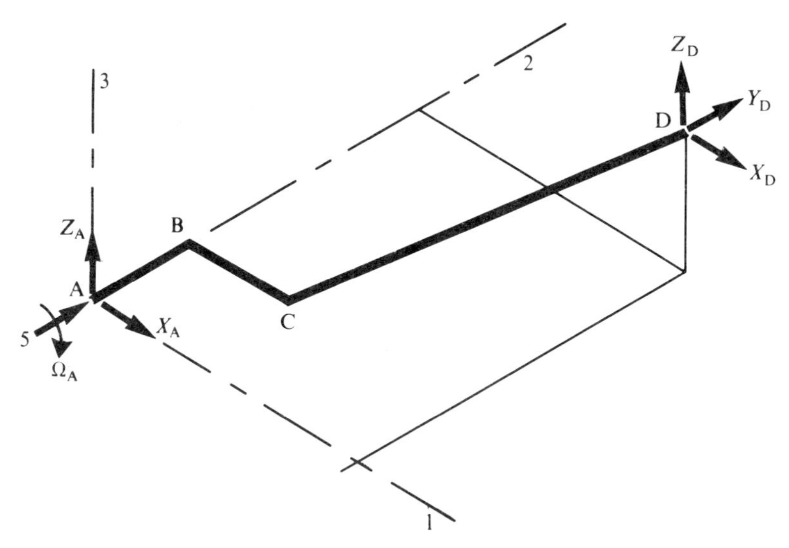

FIG. 2.12. Section of pipe.

the pipe is supported by bellows which are completely flexible in the local axial direction, so that the force they transmit to the pipe wall has no axial component, and the axial component of the force transmitted to the pipe and its contents is just the local internal pressure multiplied by the cross-sectional area; this axial force at A is 5 kN. The bellows joint can transmit torque but not bending, so that the moment exerted by the support on A is in the direction AB. At D the pipe is supported by a sealed flexible ball joint, which can transmit forces in any direction, but not moments. The pipe and the liquid within it together weigh 0·6 kN/m. What forces and moments act on the pipe and its contents at A and D?

The pipe and its contents are treated together. As far as over-all equilibrium is concerned, the weight of each segment can be treated as a concentrated force acting at its midpoint:

segment AB: weight $-1{\cdot}2\mathbf{k}$ at $0\mathbf{i}+1\mathbf{j}+0\mathbf{k}$
segment BC: weight $-1{\cdot}2\mathbf{k}$ at $1\mathbf{i}+2\mathbf{j}+0\mathbf{k}$
segment CD: weight $-3{\cdot}9\mathbf{k}$ at $3{\cdot}8\mathbf{i}+4{\cdot}4\mathbf{j}+1{\cdot}25\mathbf{k}$.

The force acting on the pipe and its contents at A is $X_A\mathbf{i}+5\mathbf{j}+Z_A\mathbf{k}$, where X_A and Z_A are the unknown 1- and 3-components, and the corresponding force at D is $X_D\mathbf{i}+Y_D\mathbf{j}+Z_D\mathbf{k}$. At A the moment acting on the pipe has an unknown magnitude Ω_A but a known direction, that of the 2-axis, and so the moment at A is $\Omega_A\mathbf{j}$.

Since the vector sum of the forces on the pipe is zero

$$(X_A\mathbf{i}+5\mathbf{j}+Z_A\mathbf{k})+(X_D\mathbf{i}+Y_D\mathbf{j}+Z_D\mathbf{k})-6{\cdot}3\mathbf{k}=\mathbf{0}, \qquad (2.3.19)$$

and so

$$X_A + X_D = 0$$
$$Y_D = -5 \qquad (2.3.20)$$
$$Z_A + Z_D = 6{\cdot}3.$$

Taking moments about the origin A

$$\mathbf{0} = \{1\mathbf{j}\times(-1{\cdot}2\mathbf{k})\} + \{(1\mathbf{i}+2\mathbf{j})\times -1{\cdot}2\mathbf{k}\} + \{(3{\cdot}8\mathbf{i}+4{\cdot}4\mathbf{j}+1{\cdot}25\mathbf{k})\times(-3{\cdot}9\mathbf{k})\}$$
$$+\{(5{\cdot}6\mathbf{i}+6{\cdot}8\mathbf{j}+2{\cdot}5\mathbf{k})\times(X_D\mathbf{i}-5\mathbf{j}+Z_D\mathbf{k})\}+\Omega_A\mathbf{j} \qquad (2.3.21)$$
$$= \mathbf{i}(6{\cdot}8Z_D - 8{\cdot}26) + \mathbf{j}(2{\cdot}5X_D - 5{\cdot}6Z_D + \Omega_A + 16{\cdot}02) + \mathbf{k}(-6{\cdot}8X_D - 28), \qquad (2.3.22)$$

and so

$$0 = 6{\cdot}8Z_D - 8{\cdot}26$$
$$0 = 2{\cdot}5X_D - 5{\cdot}6Z_D + \Omega_A + 16{\cdot}02 \qquad (2.3.23)$$
$$0 = 6{\cdot}8X_D + 28,$$

and the six equations in (2.3.20) and (2.3.23) can be solved for the six unknowns;

$$X_A = 4{\cdot}12\ \text{kN}$$
$$Z_A = 5{\cdot}09\ \text{kN}$$
$$\Omega_A = 1{\cdot}08\ \text{kN m}$$
$$X_D = -4{\cdot}12\ \text{kN}$$
$$Y_D = -5\ \text{kN}$$
$$Z_D = 1{\cdot}21\ \text{kN}.$$

This example is taken further in Example 4.3.1 (page 60).

2.4. Problems

1. A beam 8 m long rests on two supports, one at one end of the beam and the other 3 m from the other end (Figure 2.13). A vertical concentrated load of 60 kN rests on the beam at the centre. What are the reactions at the supports? The weight of the beam itself is negligible.

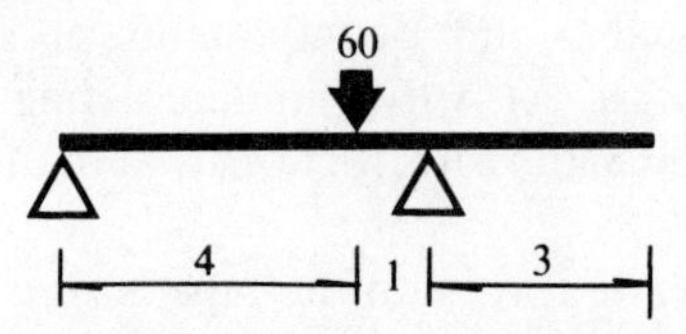

dimensions in metres; load in kilonewtons

FIG. 2.13.

2. Instead of the 60 kN concentrated load, the beam considered in Problem 1 carries a 60 kN load uniformly distributed over the left-hand half of the beam. What are the support reactions? If an upward concentrated load is added at the right-hand end, how large can it be before the reaction at the right-hand support falls to 10 kN?

3. A pitched-roof plane portal frame (Figure 2.14) is loaded in the way shown. What are the vertical components of the support reactions at the hinged supports A and B? What can you say about the horizontal components?

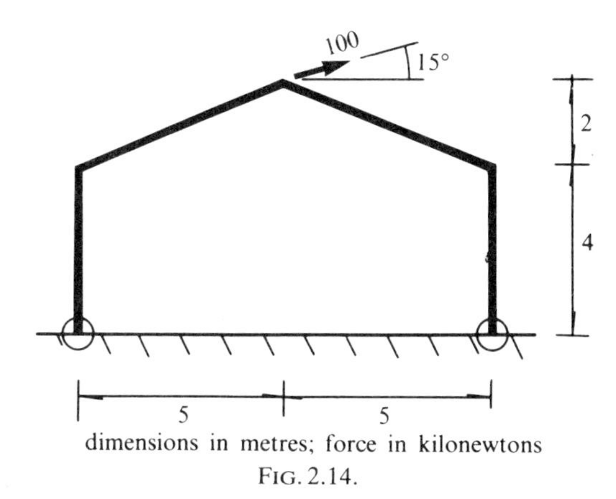

FIG. 2.14.

4. A simple flying buttress consists of two blocks of stone, ABEF and BCDE in Figure 2.15. At CD they rest on a rigid foundation. The two blocks are not connected together, or to the foundation: the joints are 'dry', without any cement.

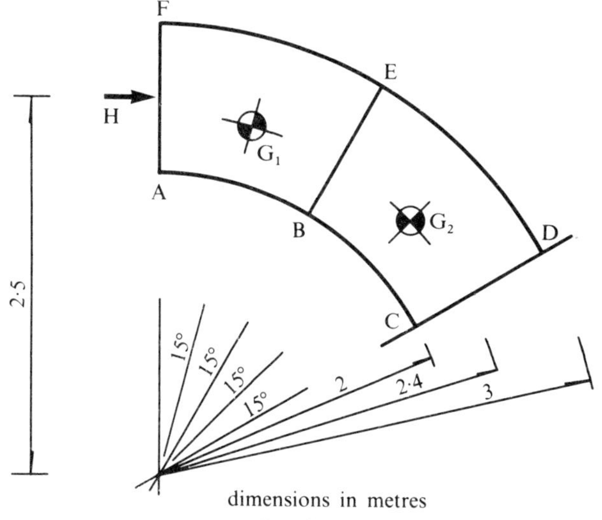

FIG. 2.15.

The blocks each weigh 30 kN. The positions of their centres of gravity G_1 and G_2 are shown in the Figure. The buttress resists a horizontal thrust H, whose line of action is shown. Find upper and lower limits on the value H can have if the buttress blocks are to be in equilibrium. Assume that the joints BE and CD are rough enough for sliding not to occur. Is this assumption reasonable?

5. A 32 m long box girder is to be launched across a canal by the method shown in Figure 2.16. It has been rolled to the left bank. When it is finally in position, a bearing on the girder at A_1 will be at A_2 on the left pier, and a bearing at B_1 will be at B_2. The girder is supported and pulled to the right by a cable which runs from B_1 over a fixed pulley at E to a winch. A second cable runs horizontally from A_1 to a holdback winch, and prevents the girder from running forward out of control. The second cable is slowly paid out as the girder advances. The lower flange of the girder runs on rollers at A_2.

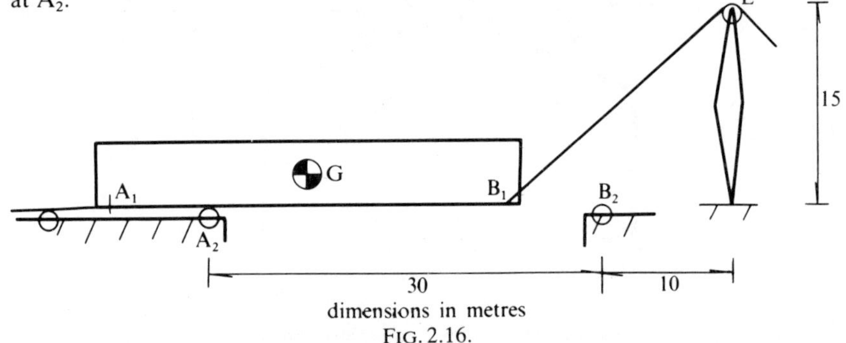

FIG. 2.16.

The girder weighs 900 kN, and its centre of gravity is at G, which is 2 m above the midpoint of A_1B_1. The engineer in charge of the launching operation wishes to know how the tensions in the two cables can be expected to change as the girder moves across the gap. Prepare a graph telling him what to expect. Assume first that the rollers at A_2 are frictionless. If instead the rollers do not move freely, but behave as a rough support, so that there is a horizontal reaction at A_2 opposing motion and equal to 0·1 times the vertical reaction, how much difference does this make?

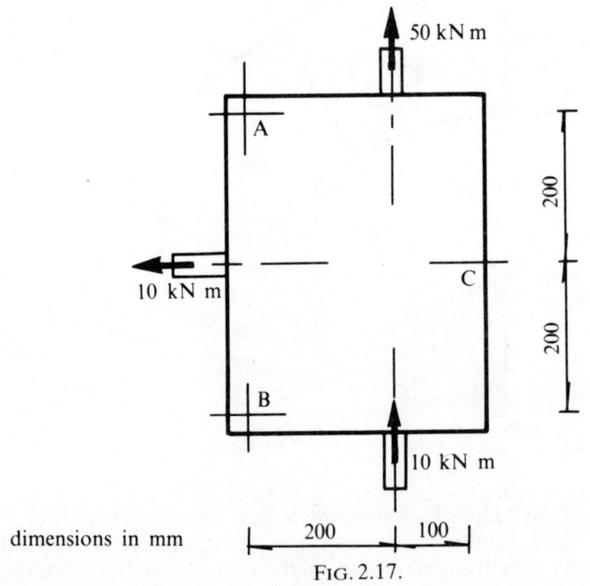

FIG. 2.17.

6. The rear-axle gear box of a vehicle, shown in plan in Figure 2.17, connects an input drive shaft 1 to two output shafts 2 and 3, which drive the rear wheels. At a certain instant the torques exerted on the shafts are 10 kN m, 50 kN m, and 10 kN m, in the directions shown. The gear box is held to the vehicle by three bolts A, B, and C. The three shafts, and the attachment points of the three bolts, all lie in the same plane.

By considering the equilibrium of the gearbox, determine the reactions which the bolts must apply. Neglect the weight of the gearbox itself.

7. A ramp in a multi-storey car park is in the form of half a turn of a helicoid (Figure 2.18). Its outer edge is a helix of radius 10 m and pitch 3 m, and its inner edge a helix of radius 6 m and the same pitch. The ramp is supported by concentrated forces at the ends A (coordinates 0, −8, 0) and C(0, 8, 3), and by a column at B (6, 0, 1·5). At A the support can exert a reaction in any direction. At C the reaction is parallel to the 1, 3-plane, and has no 2-component. At B the reaction exerted by the column is vertical.

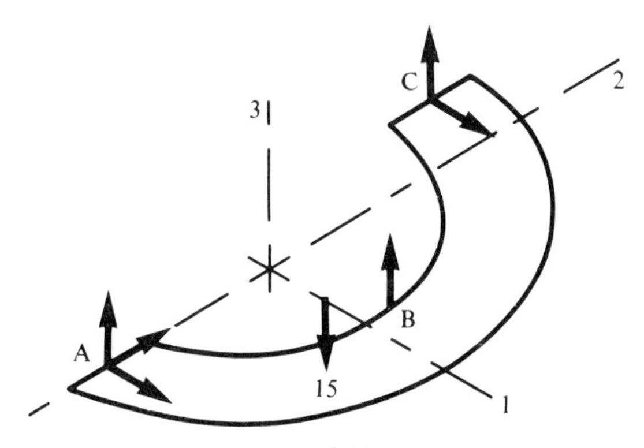

FIG. 2.18.

A car weighing 15 kN is one third of the way up the ramp, and the load it imposes can be considered as a concentrated vertical force at $(4\sqrt{3}, -4, 1)$. Calculate the corresponding support reaction components at A, B, and C.

Estimate the speed at which a car might be driven up such a ramp, and decide whether or not the corresponding centrifugal force has a significant effect on the support reactions.

3. Internal forces in structures: pin-jointed frameworks

3.1. Pin-jointed frameworks

IN the previous section we examined the forces that keep a complete structure in equilibrium. Now we look within the structure, to find what forces are set up in its different elements by the loads acting on it from outside. Just as the whole structure is in equilibrium, so must be each of its constituent parts, and the equilibrium conditions stated in Section 2.2 will apply to any part of the structure, whether large or small. We can therefore use the conditions to find forces within structures.

Chapter 1 discussed the idea of a structural idealization, an abstract version of a real structure which represents it in some kind of simplified and idealized way that lends itself to analysis. One kind of idealization has particularly simple internal forces, and we shall analyse it first. This is the *pin-jointed framework*, a framework made up of a number of straight weightless bars freely hinged together at their ends, and loaded only at the joints. Figure 3.1 shows how four simple frameworks are described schematically. The lines represent bars. They meet at joints represented by points, with circles drawn round the points to indicate that the joints are freely hinged, and transmit no moments to the end of the bars. Joints not marked with circles, as in Figure 3.1, are supposed to be able to transmit bending moments. Arrows indicate external loads. Framework (a) is a triangle of three bars hinged together. Framework (b) has five bars, in the same plane, hinged to a rigid foundation on the left; the diagonals cross without being connected together. Framework (c) is three-dimensional, and has 42 bars. These three frameworks are pin-jointed. Structure (d), on the other hand, is not: it has rigid joints, and the joints connecting the lower ends of the vertical stanchions to the foundation can transmit moments. It will be useful to have a consistent convention for representing structural idealizations in diagrams. Part of this convention has been outlined above, and some further schematic representations are indicated in Figure 3.1(e)–(g). Most structures rest on a foundation, and we need to distinguish *built-in supports* (e), which can transmit forces and moments, *simple supports* (f), which can transmit forces in any direction, but not moments, and *roller supports* (g), which are free to move in one or more directions, so that one or more components of the reaction force are known to be zero.

Can real structures sensibly be idealized as pin-jointed frameworks? Often they can, and, perhaps surprisingly, this can be so even though the joints of the real structure are not hinged at all. It turns out in general that if a

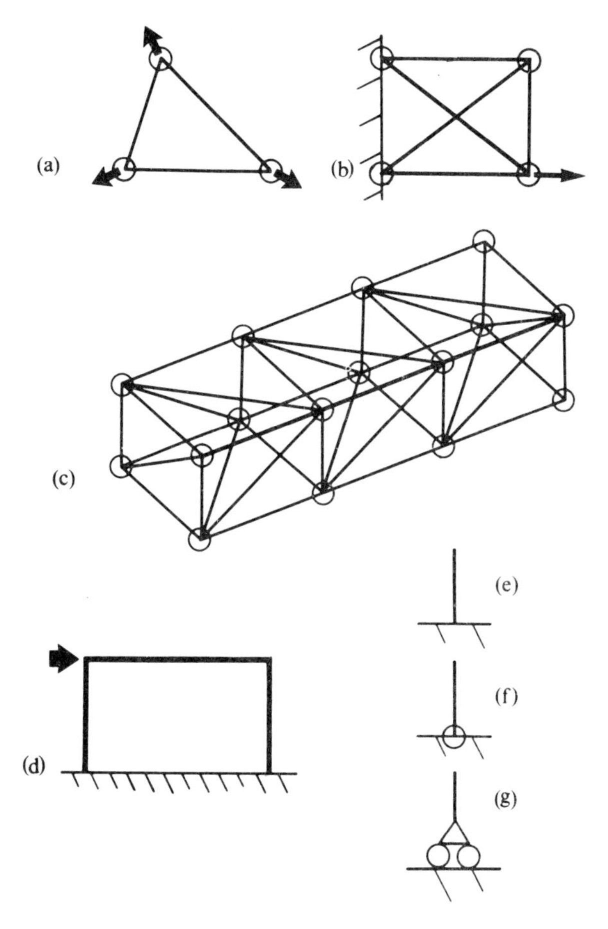

FIG. 3.1. Frameworks of different kinds.

skeletal structure would be in equilibrium if all its joints were made into free hinges, and if its bars were flexible in bending, so that they could transmit tensile or compressive axial forces but not moments, then it is reasonable to determine the primary forces in its bars by analysing it as a pin-jointed framework. If in reality the joints are not freely hinged, additional forces and moments occur, but for structures of ordinary proportions they are usually small by comparison with the primary forces. A structure like framework (d) in Figure 3.1, on the other hand, could not be analysed as pin-jointed: if its joints were made into hinges, it would become a mechanism, and it would fall over sideways.

Figure 3.2 shows an electricity pylon. It is made from long slender steel angle sections, connected together by joints which have little ability to

FIG. 3.2. Electricity pylon.

transmit bending moments. Here we would expect the pin-jointed framework idealization to be an appropriate one, and so it is. In contrast, Figure 3.3 is another view of the stern ramp in Figure 1.1. It is made from steel tubes welded together without any attempt to create a flexible hinge. The joints could carry very large moments, but because the elements of the framework are so arranged that it would be in equilibrium with hinged joints, an exact calculation shows that the member forces are much as they would be in the corresponding pin-jointed framework. Even though the framework is not pin-jointed, the pin-jointed framework idealization is a very useful one.

Each bar of a pin-jointed framework has two ends, one at each of the joints it connects. Since its own weight is neglected, it is subjected to just two forces, one exerted on the bar by the joint at one end, the other exerted by the joint at the other end. The bar is in equilibrium. In Section 2.2 we saw that if a body is in equilibrium under the action of only two forces, the forces must be equal in magnitude, opposite in direction, and in line with each other. This condition has to be obeyed here. Since the bar is straight, the end forces must be in line with the bar, as in Figure 3.4(a). If the forces were not in line, as in Figure 3.4(b), the bar could not be in equilibrium; this can be confirmed by taking moments about one end.

FIG. 3.3. Stern ramp.

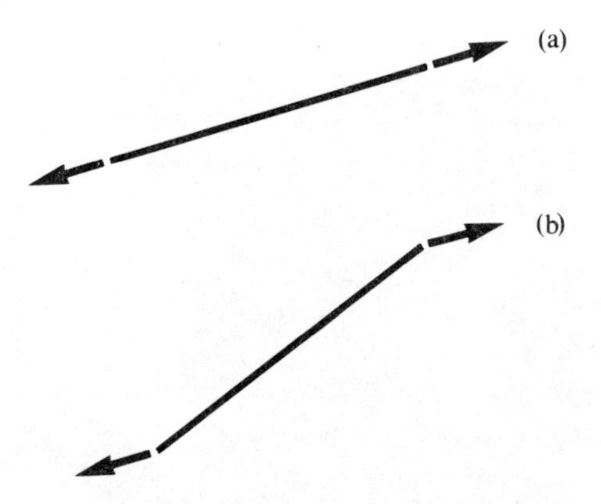

FIG. 3.4. Bar in equilibrium under end forces.

With each bar we have associated a force in line with the bar. We give the magnitude of the force a special name, calling it the *tension*, and denoting the tension in the bar joining joint I to joint J by T_{ij} (not written in bold-face type, because it is a scalar magnitude rather than a vector quantity). Calling this magnitude a tension is not meant to imply that the force is necessarily tensile. Compressive forces are denoted by negative values of the tension.

Just as each bar is in equilibrium, so is each joint. Each of the bars meeting at a joint has a tension. The vector sum of the forces corresponding to the tensions, together with any external loads acting at the joint, must be zero, since otherwise the joint could not be in equilibrium. In Figure 3.5(a), which

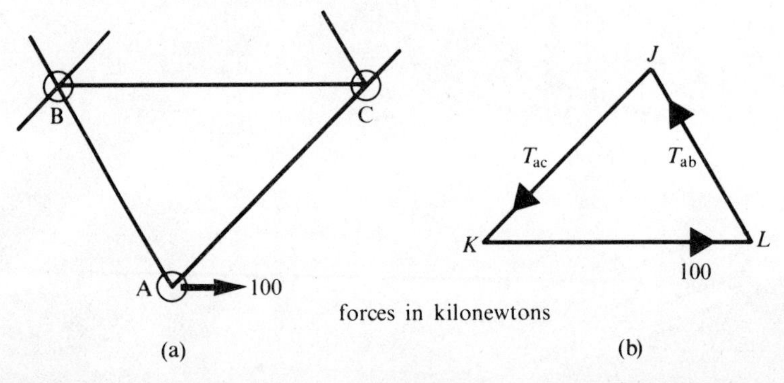

FIG. 3.5. (a) Bars connected together in a pin-jointed framework. (b) Triangle of forces for joint A.

shows part of a larger framework, two bars AB and AC meet at a joint A, where there is an external horizontal load of 100 kN. The vector sum of the forces on the joint must be zero. Figure 3.5(b) represents the corresponding vectors graphically: they must form a closed triangle of forces, with the 100 kN load represented by the horizontal vector *KL*, the axial force exerted on the joint by bar AB represented by a vector *LJ* whose magnitude is T_{ab} (and whose direction is that of the bar AB), and the tension in AC represented by a vector *JK* whose magnitude is T_{ac} (and whose direction is that of the bar AC). The directions of the forces on the joint are indicated by the heavy arrows in Figure 3.5(b). The vector *JK* acts downwards and to the left: comparison with Figure 3.5(a) shows that this must correspond to a compressive force in AC, which is what we would expect. Accordingly, T_{ac} is negative. Similarly, the vector *LJ* acts upward and to the left, and so bar AB must be in tension, and T_{ab} is positive.

In many frameworks we can find the forces in the bars by repeated application of the condition for joint equilibrium, moving on from one joint to the next. The following example illustrates how this is done.

Example 3.1.1. A square pin-jointed plane framework ABCD, attached to a rigid foundation CD, is loaded as shown (Figure 3.6). What are the tensions set up in its bars?

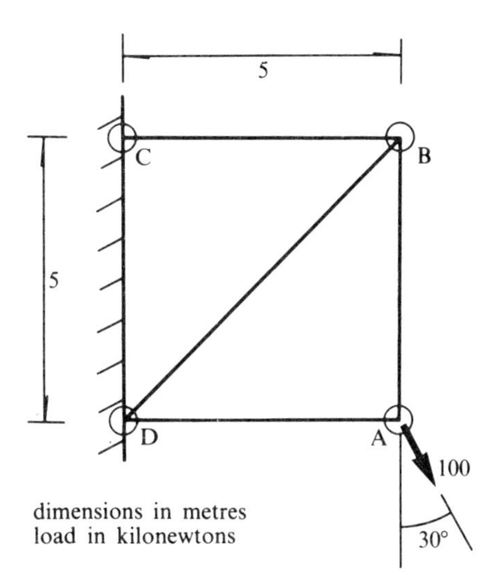

FIG. 3.6. Pin-jointed framework.

Consider joint A first. It carries the external load, and two bars meet there; their tensions are not yet known, but the directions of the corresponding forces are known, because they coincide with the directions of the bars. The triangle of forces expressing the equilibrium of joint A can be constructed, because the magnitude of one side (corresponding to the 100 kN load) and the directions of all three sides are known. It is Figure 3.7(a). By

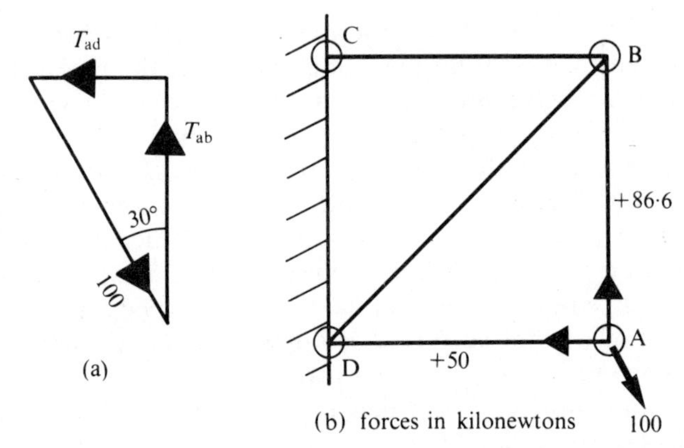

FIG. 3.7. (a) Triangle of forces for joint A. (b) Framework, showing tensions in bars AB and AD.

trigonometry, the magnitudes of T_{ab} and T_{ad} are respectively 100 cos 30° = 86·6 and 100 sin 30° = 50. Comparison with Figure 3.6 shows that AB and AD are both in tension, and so

$$T_{ab} = +86{\cdot}6 \text{ kN}$$

$$T_{ad} = +50 \text{ kN}.$$

It is often helpful to write the tensions on the framework layout itself. A tension, with its sign, is written next to the centre of the coresponding bar. In addition, two arrows are drawn on the bar, one at each end. An arrow indicates the direction of the force exerted on the neighbouring joint by that end of the bar. In this instance, both T_{ab} and T_{ad} are positive, and so arrows are drawn on AB and AD, pointing away from A. At the other end of AB,

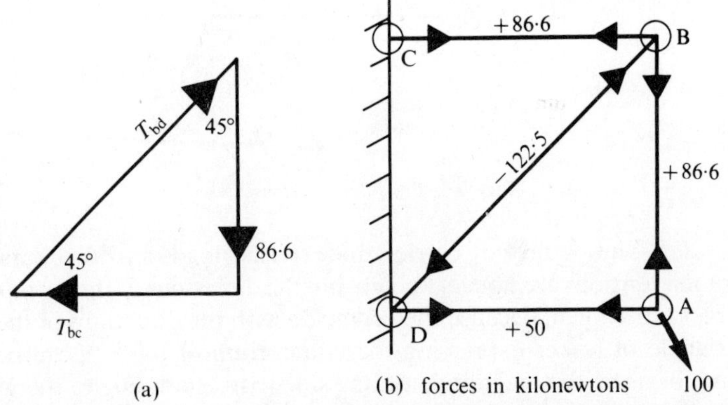

FIG. 3.8. (a) Triangle of forces for joint B. (b) Framework, showing tensions in all four bars.

the force exerted by bar AB on joint B is directed downwards, and so a second arrow is drawn, pointing towards A (Figure 3.8(b)). Note that there is a duplication of information here, in that the sign of the tension is indicated both by the arrows (which point outward if the bar is in compression, inward if it is in tension) and explicitly by the sign of the numerical value.

We now move on to joint B. Three bars meet here. The directions of the three forces on the joint are known, and one of the magnitudes, T_{ab}, was calculated in the previous step, and so the triangle of forces for joint B can be drawn (Figure 3.8(a)), and solved by trigonometry. Comparison between the triangle of forces and the structure layout shows that BC is in tension and BD in compression. Therefore

$$T_{bc} = +86{\cdot}6 \text{ kN}$$

$$T_{bd} = -86{\cdot}6/\cos 45° = -122{\cdot}5 \text{ kN},$$

and the bar tensions are all known (Figure 3.8(b)). A check could be made by calculating the external reactions on the framework at C and D, and confirming that they balance the 100 kN load, so that the whole framework is in equilibrium.

Notice that joint A was considered first, and then joint B. It would not have been possible to work the other way round, because at the start of the calculation none of the tensions in the three bars that meet at B has a known magnitude.

Example 3.1.2. A plane pin-jointed truss rests on a simple support at A and a roller support at E (Figure 3.9). The loading is shown: what are the tensions in the bars of the framework?

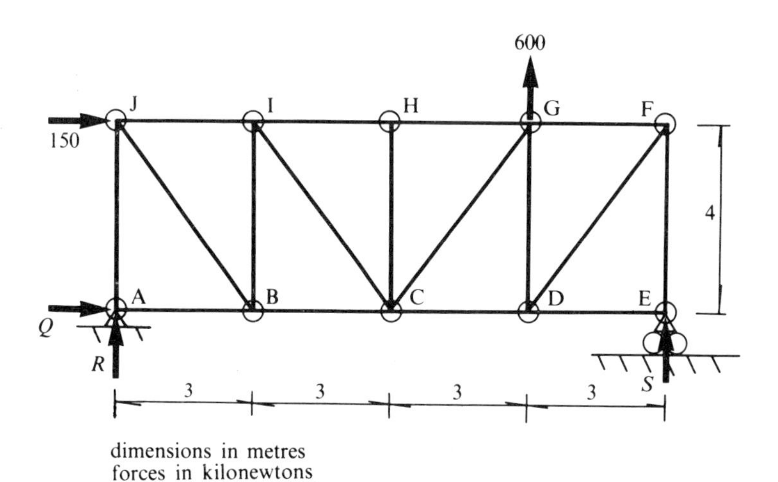

FIG. 3.9. Framework.

One has to decide which joint to apply the equilibrium conditions to first. By analogy with the previous example, the obvious choice is one of the loaded joints G and J. At G, however, there are four unknown bar tensions, too many for it yet to be possible to construct the closed force polygon expressing the equilibrium of the joint. This is also the case at J. Instead, begin by finding the support reactions Q, R, and S, by the methods of Section 2.3. Then

$$Q = -150 \text{ kN}$$

$$R = -400 \text{ kN}$$

$$S = -200 \text{ kN}.$$

It is then easy to start the construction of triangles of forces at either A or E, at either of which there are only two unknown tensions. From then on the method of solution is exactly as in Example 3.1.1. One works from joint to joint, at each step choosing as the next joint one which only has two bars with

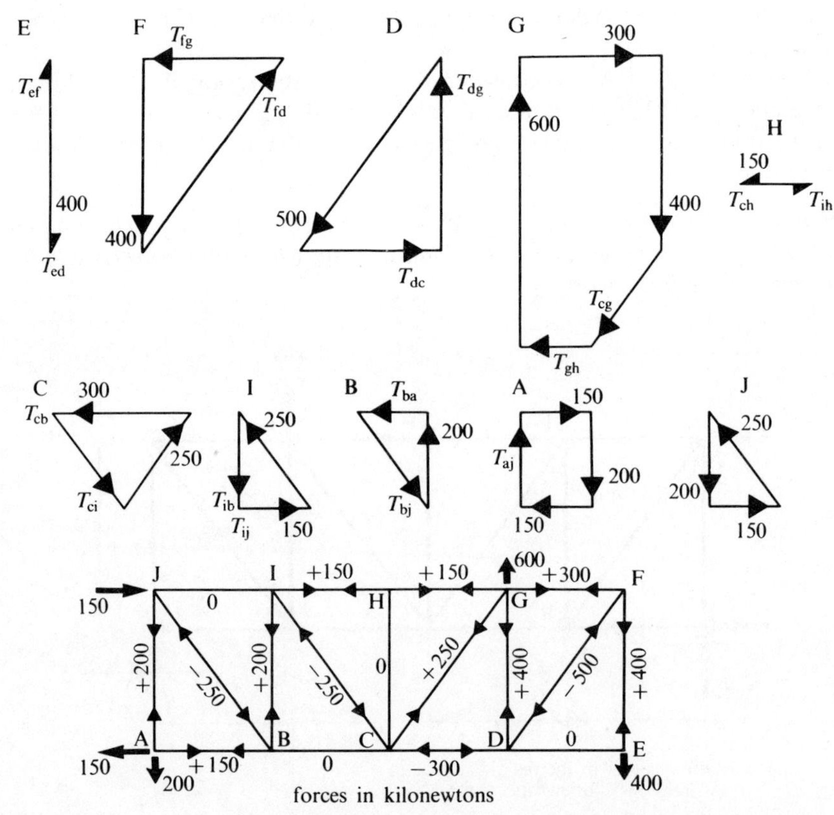

FIG. 3.10. Polygons of forces for the joints of the framework in Figure 3.9.

unknown tensions. Starting at E, the joint sequence is E F D G H C I B A J (the last only being needed as a check). The corresponding force polygons are shown in order in Figure 3.10, together with the final solution listing all the tensions. At each step in the force polygon sequence, tensions known at the start of that step are given their numerical values, and unknowns are indicated by symbols. You are encouraged to work through these polygons in sequence, checking the directions and magnitudes of the various vectors.

In this example we worked through the structure by a particular route from joint to joint, and it turned out that there were never too many unknown forces at a joint. Obviously the method is not completely systematic. In addition, there do exist unusual framework layouts for which this method fails, because there is no joint at which there is enough information to start the analysis. An alternative method is simply to write down all the equilibrium equations for all the joints. There are 17 bars, and therefore 17 unknown bar tensions, and three reactions at the supports, making 20 unknown quantities in all. At each joint the vector sum of the loads and bar forces is zero, and so the vertical and horizontal components of the vector sum are individually zero. There are 10 joints, and therefore 20 equations for the 20 unknowns, and the equations can be solved for the unknowns. The equations are

$$
\begin{aligned}
T_{ab}+Q&=0 && \text{A, horizontal}\\
T_{aj}+R&=0 && \text{A, vertical}\\
T_{bc}-0{\cdot}6T_{bj}-T_{ab}&=0 && \text{B, horizontal}\\
T_{bi}+0{\cdot}8T_{bj}&=0 && \text{B, vertical}\\
T_{cd}+0{\cdot}6T_{cg}-0{\cdot}6T_{ci}-T_{cb}&=0 && \text{C, horizontal}\\
0{\cdot}8T_{cg}+T_{ch}+0{\cdot}8T_{ci}&=0 && \text{C, vertical}\\
T_{de}+0{\cdot}6T_{df}-T_{cd}&=0 && \text{D, horizontal}\\
0{\cdot}8T_{df}+T_{dg}&=0 && \text{D, vertical}\\
T_{de}&=0 && \text{E, horizontal}\\
T_{ef}+S&=0 && \text{E, vertical}\\
0{\cdot}6T_{df}+T_{fg}&=0 && \text{F, horizontal}\\
-0{\cdot}8T_{df}-T_{ef}&=0 && \text{F, vertical}\\
T_{fg}-0{\cdot}6T_{cg}-T_{gh}&=0 && \text{G, horizontal}\\
-T_{dg}-0{\cdot}8T_{cg}+600&=0 && \text{G, vertical}\\
T_{gh}-T_{hi}&=0 && \text{H, horizontal}
\end{aligned}
\tag{3.1.1}
$$

$$-T_{ch}=0 \qquad \text{H, vertical}$$

$$T_{hi}+0{\cdot}6T_{ci}-T_{ij}=0 \qquad \text{I, horizontal}$$

$$-T_{bi}-0{\cdot}8T_{ci}=0 \qquad \text{I, vertical}$$

$$T_{ij}+0{\cdot}6T_{bj}+150=0 \qquad \text{J, horizontal}$$

$$-0{\cdot}8T_{bj}-T_{aj}=0 \qquad \text{J, vertical}$$

Force components have been counted positive to the right and upward.

This method is more systematic, and works for any statically determinate framework, whether plane or three-dimensional, but is plainly relatively laborious. However, because the equations are set up systematically, and can be solved straightforwardly, this method is well suited to computer solutions for the forces in frameworks. Programs that carry out such solutions are nowadays widely available.

Bow's notation

Figure 3.10 shows the polygons of forces for the equilibrium of the joints of the truss in Example 3.1.2. Each of the bar tensions appears twice in this diagram, once for each of the bars it connects. Tension T_{cg}, for instance, appears both in polygon G and in polygon C (where it is given its numerical value of +250). It is natural to wonder if it might be possible to combine all these polygons in one diagram, in which each line representing a bar tension would only appear once.

There are two difficulties. First, if all the information originally given in several polygons, one for each joint, is put into a single diagram, that diagram is necessarily more complicated, and we shall need a reliable and systematic way of referring to forces within it, since otherwise it will be too involved to understand. Secondly, if there are more than three forces, the force polygons can be drawn in more than one way (depending on the order in which the forces are taken), and do not automatically fit together correctly.

An ingenious notation due to Bow overcomes both difficulties. Consider a joint K in equilibrium under four bar tensions (Figure 3.11(a)), of which three (T_{kl}, T_{km}, and T_{kn}) are positive, and the fourth, T_{ko}, is negative. Bow's notation labels the spaces between the bars, *a b c d* and so on. Lower-case letters are used, to distinguish the labelling of spaces from the labelling of joints, which uses upper-case letters. Each tension is directed along a line which separates two spaces, and we can refer to it by the two letters that label the two spaces. Thus, T_{ko} lies between space *a* and space *b*, and so we call it *ab*.

In order to construct the force polygon expressing the equilibrium of joint K, we go round the joint systematically, from one joint to the next, in the

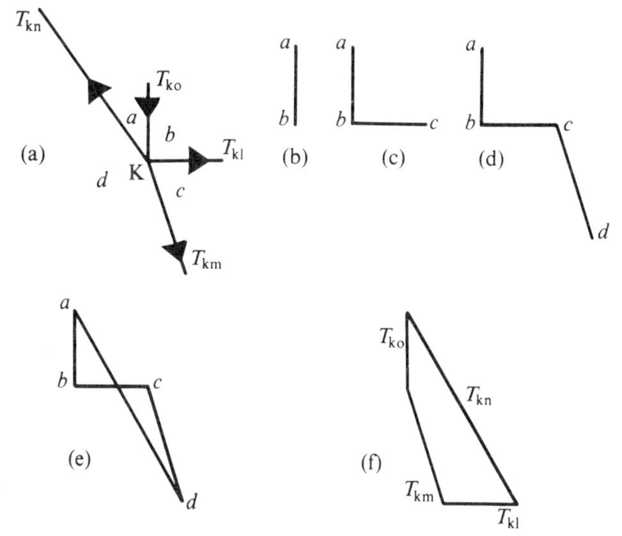

FIG. 3.11. Bow's notation.

same direction. Suppose, for example, that we start at *a* and move clockwise. Steps in the construction of the force polygon are shown in Figure 3.11(b–e). The force corresponding to T_{ko} acts downward on K; it is represented by a vector *ab*, drawn downwards from *a* to *b*, whose length corresponds to the absolute value of T_{ko} (Figure 3.11(b)). Going clockwise, the next bar tension is T_{kl}, which separates space *b* from space *c*, and is positive, so that the corresponding force acts to the right on K. Accordingly, draw *bc* to the right from *b*, its magnitude corresponding to T_{kl} (Figure 3.11(c)). Still going clockwise, T_{km} comes next; the corresponding force acts downward and to the right, and is represented by *cd* (Figure 3.11(d)). Finally, T_{kn} is represented by *da* (Figure 3.11(e)).

You should check for yourself that the same polygon is obtained whether the starting point is *a*, *b*, *c*, or *d*, as long as the sequence is clockwise round the joint. Anticlockwise sequences give the same polygon, but rotated through 180°. In applications of the notation, it is important to be consistent and to use either a clockwise or an anticlockwise sequence, but not to mix them up. The point about going round the joint systematically is important: the force polygon in Figure 3.11(f) is a perfectly correct expression of the equilibrium of joint K, but does not fit within the notation, because the sequence is $T_{ko}T_{km}T_{kl}T_{kn}$, which does not go round in a consistent direction.

In Figure 3.12 the truss in Example 3.1.2 is reanalysed with the help of Bow's notation. The reactions at the supports are first calculated in the usual way, and are included in the diagram. The spaces are labelled *a*, *b*, *c*, . . . *l*, *m*. The notations for joints and spaces are independent: there is no connection

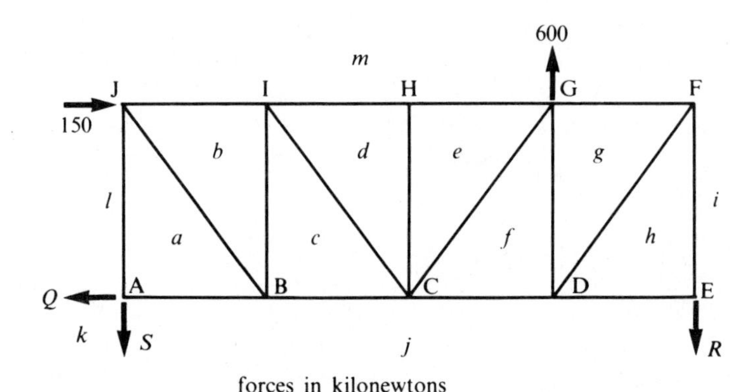

FIG. 3.12. The framework of Example 3.1.2, relabelled in accordance with Bow's notation.

between A, which identifies a joint, and *a*, which identifies the triangular region between bars BJ, JA, and AB. Spaces outside the framework are lettered as well, the dividing lines being external forces and reactions, so that *i* is the space between the external load 600 and the reaction *R*.

We can now identify forces in the same way as in Figure 3.11, by the two letters labelling the spaces on either side of them. Bar AB, for instance, separates space *a* from space *j*, and in force polygons it will be represented by a vector *aj*.

The method and the joint sequence are the same as in the previous analysis of this problem, but the notation is different, and now all the force polygons are brought together. Begin at joint E. At this joint the external vertical support reaction, now denoted *ij*, is 400 kN, acting downwards. This is represented by a vertical line *ij*, downwards from *i* to *j*, which starts the force polygon (Figure 3.13(a)). The three forces on the joint are represented by *ij*, *jh*, and *hi*. Points *i* and *j* have already been located. Force *jh* is

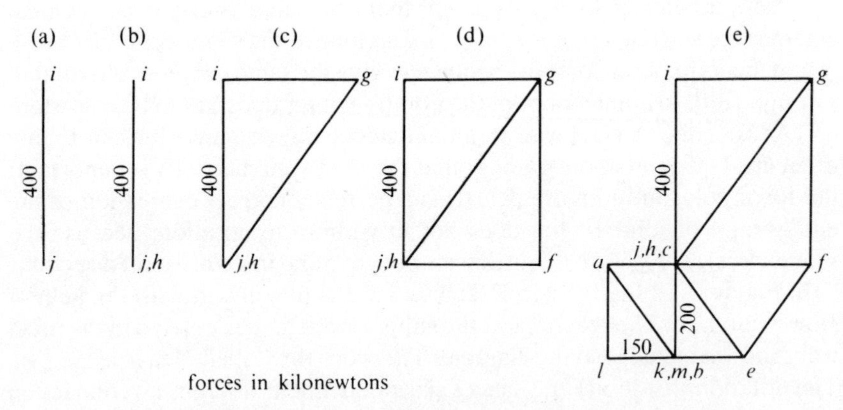

FIG. 3.13. Solution of Example 3.1.2 with the help of Bow's notation.

horizontal, and so h lies on a horizontal line through j. Force hi is vertical, and so h lies on a vertical line through i. The only point that satisfies these conditions is j itself, and so h coincides with j (Figure 3.13(b)). The tension T_{de} in bar DE is represented by hj, and is zero. The tension T_{ef} in EF is hi, upwards from h to i, and so EF is in tension and T_{ef} is +400 kN.

Continue to joint F, in equilibrium under forces ih, hg, and gi. Since gi is horizontal, g lies on a horizontal line through i. The direction of hg is parallel to bar DF, and g lies on a line in this direction through h. These two conditions together locate g (Figure 3.13(c)). The triangle of forces for F is gih: gi acts to the left, and so bar FG is in tension, while hg is upwards and to the right, and so DF is in compression.

Continue to joint D. Its polygon of forces will be $ghif$. Point f has not yet been located, but jf is horizontal, and gf is vertical, and these conditions locate f (Figure 3.13(d)). Force gf is downwards and to the left, and so DF is in compression; hj is zero; jf is to the left (and so CD is in compression); fg is upward (and so DG is in tension).

A complete force polygon can be constructed in this way. It is shown in Figure 3.13(e), which contains all the polygons for all the joints.

Bow's method emphasises the power and usefulness that an apparently trivial change in notation can have. Methods of this kind are useful elsewhere in engineering, such as in electrical circuit theory, and are linked to graph theory, which has applications in many subjects, among them architectural design and switching-network theory.

3.2. Statical indeterminacy

Example 3.2.1. A square pin-jointed plane framework ABCD (Figure 3.14(a)), attached to a rigid foundation at C and D, is loaded as shown. What are the tensions set up in the bars?

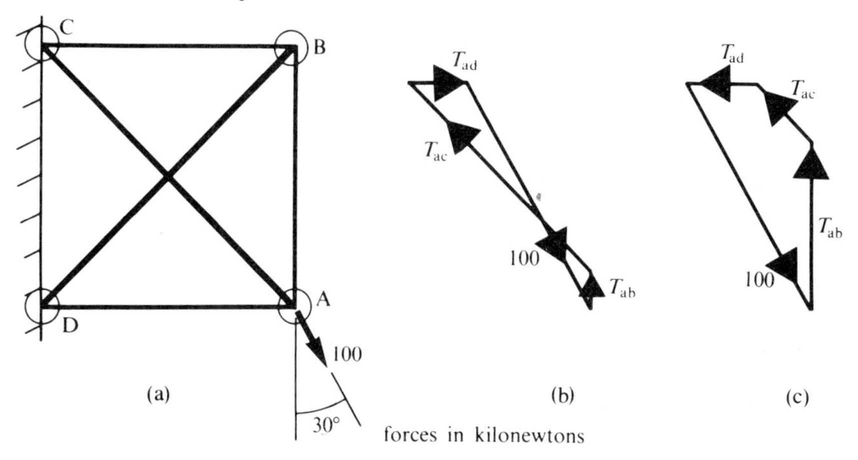

FIG. 3.14. (a) Statically indeterminate framework. (b) Possible force polygons expressing the equilibrium of joint A.

This framework is like the one analysed in Example 3.1.1, but has one extra bar, AC. In that example we began by considering the equilibrium of joint A. If we do so again here, we find that there are three unknown tensions at A, T_{ab}, T_{ac}, and T_{ad}. We know their directions, but none of their magnitudes, and so we do not have enough information to construct the force polygon expressing equilibrium of A. Two of the infinitely many possible polygons are shown in Figure 3.14(b): both have the right directions for the bar forces, and the same magnitude and direction representing the load, but they are obviously different. Both express an equilibrium condition, but fail to determine the unknown tensions.

The same thing happens if we start the analysis at B, or at the supports C and D. This is another statically indeterminate structure, as we found the structure of Example 2.3.3 to be statically indeterminate. If we try to use the alternative method, and write down and solve the joint equilibrium equations, we find that that method also fails. There are five unknown bar tensions, and four unknown support reaction components (two at C, one vertical and one horizontal, and two at D), making altogether nine unknowns. There are four joints, and two equilibrium equations for each, and therefore eight equations, not enough for us to be able to solve them for the nine unknowns.

In Chapter 6 we shall see how to find the forces in statically indeterminate frameworks.

3.3. Method of sections

It sometimes happens that we want to find one or two of the bar tensions in a large structure, without going to the trouble of using one of the methods described above to find all the tensions. It is then often useful to use a method which depends on the equilibrium conditions for part of the framework cut off by an imaginary section.

Example 3.3.1. Figure 3.15(*a*) *shows a bridge truss with* 23 *bars. It carries vertical loads of* 100 *kN at D,* 140 *kN at E, and* 140 *kN at F. What tensions are set up in bars EF, EI, and IJ*?

The support reactions are found first. They are

$$120\text{ kN at A}$$

$$260\text{ kN at G.}$$

Imagine the truss divided by an imaginary cut, indicated by the dashed line in Figure 3.15(a), which cuts the three bars whose tensions we want to find. Consider the equilibrium of the segment of the truss to the right of this cut, shown in Figure 3.15(b). Five forces act on this segment. Two of them are known, the load at F and the external reaction at G. The remaining three

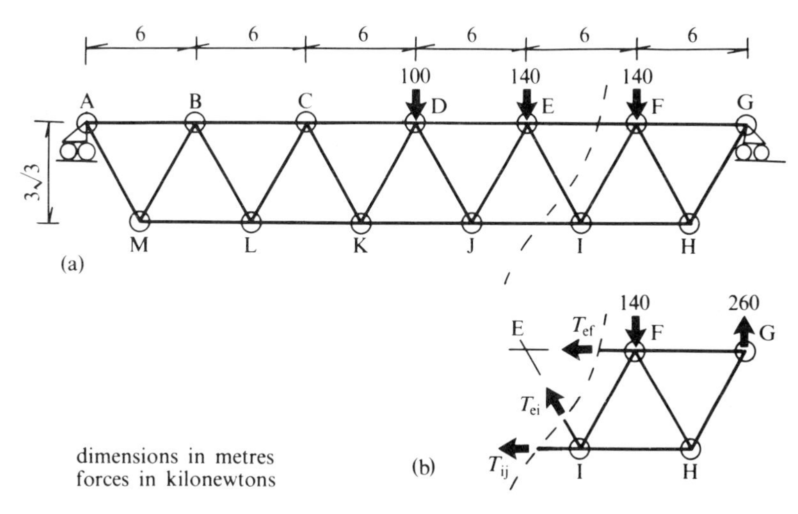

FIG. 3.15. (a) Bridge truss. (b) Segment to the right of imaginary cut across truss.

correspond to the bar tensions T_{ef}, T_{ei}, and T_{ij}. Together the five forces keep the segment in equilibrium, and so we can exploit its equilibrium conditions to find the unknown bar tensions. The resultant vertical force on the segment must be zero. The tensions in EF and IJ act horizontally, and so have no vertical component. The equilibrium condition for the segment to the right of the cut is

$$T_{ei} \sin 60° - 140 + 260 = 0 \tag{3.3.1}$$

and so

$$T_{ei} = -138 \text{ kN}.$$

The lines of action of the tensions T_{ei} and T_{ef} both pass through point E, and therefore have no moments about that point. If we take moments about E, we have an equation in which only T_{ij} appears. Since the resultant moment about E of the forces acting on the segment is zero

$$\begin{aligned} (3\sqrt{3})T_{ij} + (6)(140) - (12)(260) &= 0 \\ T_{ij} &= +439 \text{ kN}. \end{aligned} \tag{3.3.2}$$

Finally, taking moments about I, which the lines of action of T_{ei} and T_{ij} both pass through,

$$\begin{aligned} (3\sqrt{3})T_{ef} + (9)(260) - (3)(140) &= 0 \\ T_{ef} &= -370 \text{ KN}. \end{aligned} \tag{3.3.3}$$

As a final check, the resultant horizontal force on the segment must be zero, and so

$$T_{ef} + T_{ei} \cos 60° + T_{ij} = 0, \tag{3.3.4}$$

and the above results are consistent with this.

It was an arbitrary choice to use the equilibrium conditions for the segment to the right of the imaginary cut. The segment to the left would have done equally well.

3.4. Three-dimensional frameworks

Three-dimensional frameworks can be analysed by essentially the same methods as plane frameworks, but their additional complexity alters the relative importance and usefulness of the different techniques. The systematic method is to write down all the joint equilibrium equations, and then to solve them: that will always work, as long as the framework is statically determinate, and is usually the most straightforward technique. In some structures it is possible to speed up the analysis, either by using the method of sections or by working from one joint to the next, and finding all the tensions at each joint in turn, but short cuts like this are less generally useful than they are in plane frameworks.

The geometry of three-dimensional frameworks is naturally more complicated than that of plane frameworks. Figure 3.16 shows a bar AB connecting

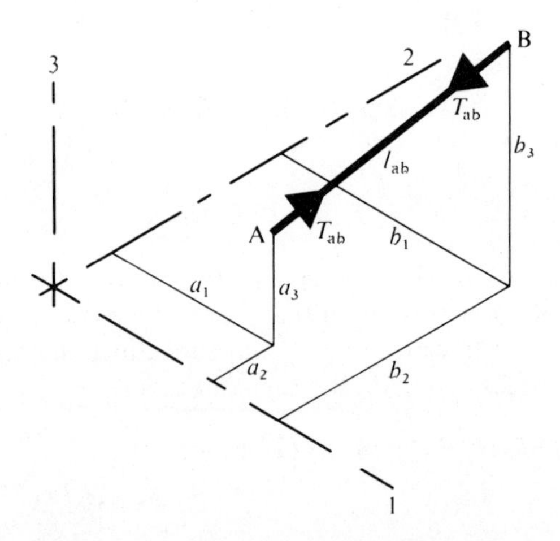

FIG. 3.16. A bar in a three-dimensional framework.

joints A (coordinates a_1, a_2, and a_3) and B (coordinates b_1, b_2, and b_3). Call the length of the bar l_{ab}, so that

$$l_{ab}^2 = (b_1 - a_1)^2 + (b_2 - a_2)^2 + (b_3 - a_3)^2.$$

The cosine of the angle between AB and the 1-axis is the projection of AB onto the axis divided by the length of AB, that is $(b_1 - a_1)/l_{\mathrm{ab}}$. The component in the 1-direction of the force exerted by AB on A is the tension T_{ab} multiplied by this cosine, that is

$$T_{\mathrm{ab}}(b_1 - a_1)/l_{\mathrm{ab}}.$$

In the same way, the 2- and 3-components are

$$T_{\mathrm{ab}}(b_2 - a_2)/l_{\mathrm{ab}}$$

and

$$T_{\mathrm{ab}}(b_3 - a_3)/l_{\mathrm{ab}}.$$

Accordingly, if we resolve forces on joints in mutually orthogonal directions, which is usually the simplest thing to do, the components we need are immediately derivable from the geometry of the framework.

Example 3.4.1. A pin-jointed framework is built up of nine bars, pinned to a rigid foundation. Its configuration and dimensions are given in Figure 3.17. A load of 50 kN acts at joint A; its line of action lies in a plane parallel to BCFE, at 15° to EF. How can the tensions in the bars be found?

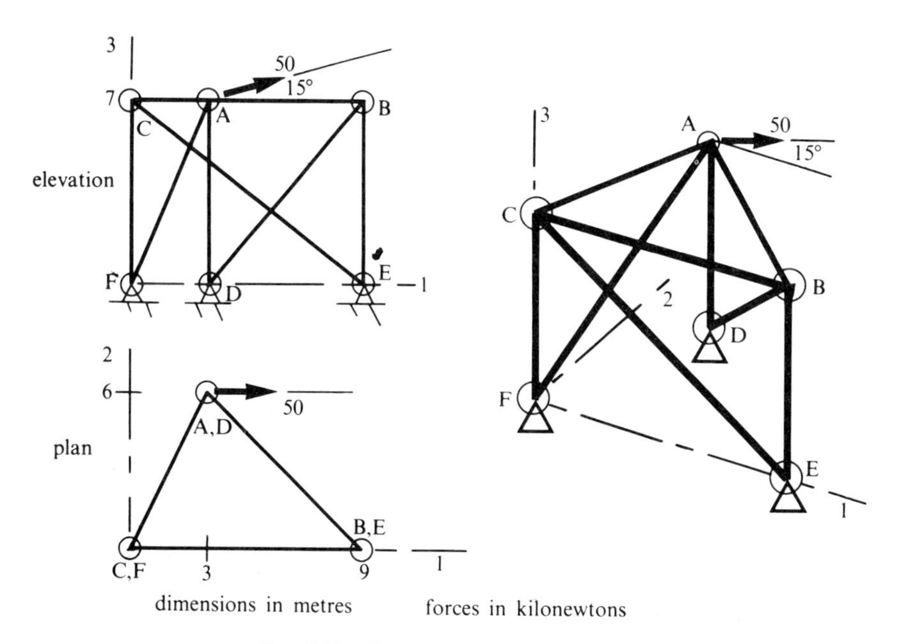

FIG. 3.17. Three-dimensional framework.

The lengths of the bars are

AB	$\sqrt{72}$	BD	11	AD	7
BC	9	CE	$\sqrt{130}$	BE	7
CA	$\sqrt{45}$	AF	$\sqrt{94}$	CF	7

and the direction cosines of the line of action of the load are cos 15°, 0, and cos 75°. At joint A, there act tensions corresponding to four bars, and the external load. Since the 1-component of the vector sum of the forces on the joint must be zero, the sum of the 1-components of the individual forces must be zero. These components are

$(6/\sqrt{72})T_{ab}$ in the positive 1-direction, from bar AB, and using the earlier result for force components,
$(3/\sqrt{45})T_{ac}$ in the negative 1-direction,
$(3/\sqrt{94})T_{af}$ in the negative 1-direction, and, from AD
$(0/7)T_{ad}$, because its projection in the 1-direction is zero.

Equilibrium therefore requires that

$$(6/\sqrt{72})T_{ab}-(3/\sqrt{45})T_{ac}-(3/\sqrt{94})T_{af}+50\cos 15°=0. \qquad (3.4.1)$$

Similarly, equating to zero the resultant force in the 2-direction

$$-(6/\sqrt{72})T_{ab}-(6/\sqrt{45})T_{ac}-(6/\sqrt{94})T_{af}=0, \qquad (3.4.2)$$

and in the 3-direction

$$-T_{ad}-(7/\sqrt{94})T_{af}+50\sin 15°=0. \qquad (3.4.3)$$

Here are three equations for four unknown quantities. In this instance we cannot find any bar forces from the equilibrium conditions for a single joint. To the three equations for joint A we can add three additional equations for joint B

$$-(6/\sqrt{72})T_{ab}-T_{bc}-(6/11)T_{bd}=0 \qquad (3.4.4)$$

$$(6/\sqrt{72})T_{ab}+(6/11)T_{bd}=0 \qquad (3.4.5)$$

$$(7/9)T_{bd}+T_{be}=0, \qquad (3.4.6)$$

and three more for joint C

$$(3/\sqrt{45})T_{ac}+T_{bc}+(9/\sqrt{130})T_{ce}=0 \qquad (3.4.7)$$

$$(6/\sqrt{45})T_{ac}=0 \qquad (3.4.8)$$

$$-(7/\sqrt{130})T_{ce}-T_{cf}=0. \qquad (3.4.9)$$

There are now nine independent equations for nine bar tensions. Because of the geometry of this particular structure, it happens that T_{ad} only appears in eqn (3.4.3), T_{be} only in eqn (3.4.6), and T_{cf} only in eqn (3.4.9). Leaving out these three equations, there remain six independent simultaneous equations in six unknowns. Their solution is straightforward, whether by hand or by computer.

You will notice that in these equations the coefficient of each bar tension has the bar length in the denominator: thus the coefficients of T_{af} are $-3/\surd 94$, $-6/\surd 94$, and $-7/\surd 94$, and $\surd 94$ is the length of AF. This is because of the relationship between bar tensions, lengths, and force components. The method of tension coefficients (Parkes (1965)) is the same method described above, but uses as its variable describing each bar a tension coefficient, the bar tension divided by its length, so that AF would be described by $T_{af}/\surd 94$. Sometimes this simplifies the arithmetic of hand calculations.

How complex the simultaneous equations are depends very much on the detailed arrangement of the bars. Figure 3.18 shows a framework very like

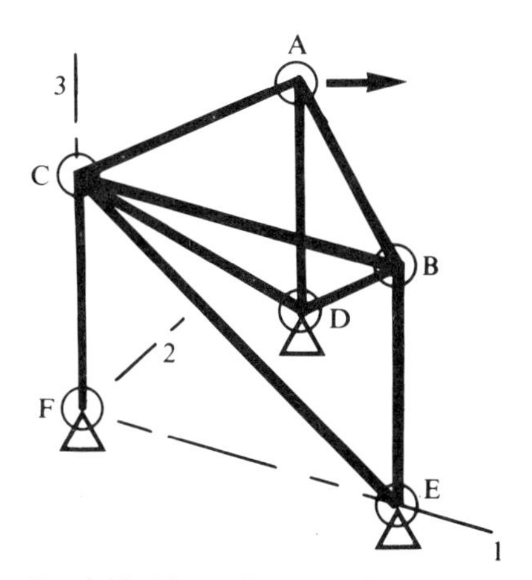

FIG. 3.18. Three-dimensional framework.

the one in Figure 3.17, the only difference being that bar AF has been removed and a bar CD added. This minor change makes the solution much easier. For the new structure, the equilibrium equations for A, corresponding to eqns (3.4.1) to (3.4.3), now only have three unknowns, and can be solved immediately for T_{ac}, T_{ab}, and T_{ad}. The equations for B then give T_{bd}, T_{bc}, and T_{be}, and those for C give T_{cd}, T_{ce}, and T_{cf}.

3.5. Under what conditions is a framework statically determinate?

It is easiest first to think about systems of bars freely hinged together in such a way that all the bars are in the same plane. Figure 3.19 shows a number of such systems. You may find it helpful to make models of them, using a construction toy such as Meccano, or strips of card or paper hinged

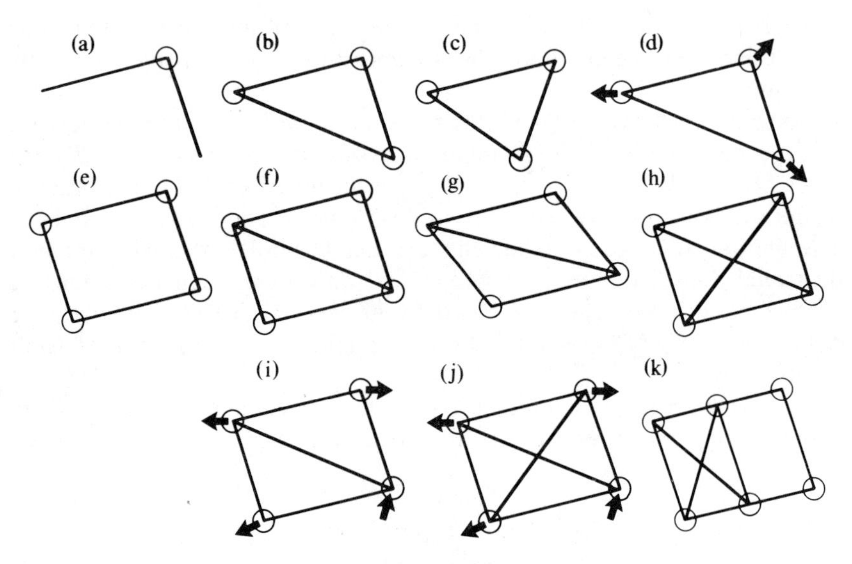

FIG. 3.19. Plane pin-jointed frameworks.

together by drawing pins (thumb-tacks) or paper fasteners. System (a) is not a structure but a mechanism: the angle between the bars can be altered freely. If we add a third bar, to make (b), the configuration becomes fixed: the length of the third bar determines the geometry, and the system has become a rigid structure, which cannot distort unless the lengths of the bars change. We say that it is kinematically determinate, in contrast to (a), which is kinematically indeterminate. This is so whatever the length of the third bar, whether long as in (b) or short as in (c). Imagine now that we apply external forces to the joints of (b), as in (d), and the external forces are such that the whole framework is in equilibrium. The induced bar tensions can be found by considering the equilibrium of the joints. It turns out that there is just the right amount of information for this to be possible, using the method described earlier in this section. As well as being kinematically determinate, the framework is statically determinate. If we look at it in terms of equilibrium equations, got by equating to zero the resultant forces in two perpendicular directions on each of the three joints, there are six such equations. However, because the external loads have to obey three equilibrium conditions, only three of the joint equilibrium equations are independent, and this is exactly what we need to determine the three bar tensions. In system (e), four bars are hinged together: this is another mechanism, and can freely be distorted into different quadrilaterals with the same sides. A diagonal fifth bar fixes the configuration, as in (f) or (g), and makes it kinematically determinate. If now we attempt to add a second diagonal bar, to make (g) into (h), we find that it will only fit if it has exactly the right length. Once again, imagine that we apply to the joints of frameworks (e) to (h) a system of forces which is itself in equilibrium. Mechanism (e) will not generally be able to withstand these forces, unless they obey an additional condition over and above the three needed for over-all equilibrium. System (f) is a structure, and will be able to sustain the forces, as in (i), and the corresponding bar tensions can be found from equilibrium. Counting equilibrium equations for (i), there are eight joint equations, but three over-all equilibrium conditions, and so five independent joint equilibrium equations, exactly what we need to find the unknown tensions induced in the five bars. The framework is statically determinate. However, if we apply forces to (h), as in (j), we find that it is no longer possible to find the bar tensions by statics. Like the structure in Example 3.2.1, (h) is statically indeterminate. There are still only five independent joint equilibrium equations, but there are six unknown bar tensions.

It turns out that the conditions for a framework just to be kinematically determinate, in the sense that it has just enough bars for its configuration to become fixed, so that it is not a mechanism, are the same as the conditions for it to be statically determinate, in the sense that the forces within it can be found from the external loads. This is an example of a duality between statics and kinematics that we shall meet again. Most people find it easier to visualize the deformation of a framework than the forces within it. A simple way to determine whether a framework is statically determine is to think of removing a single bar from it. If it becomes a mechanism, then it was statically determinate to begin with. If not, then it was statically indeterminate, and further bars have to be removed to make it a mechanism. For instance, Figure 3.18(f) represents a statically determinate framework, because if any of the bars is removed it becomes a mechanism. If the diagonal is removed, it becomes (e); if one of the other bars is removed instead, it becomes a triangle with one 'loose' bar attached to a corner.

An alternative way of deciding whether or not a framework is statically determinate is to think of how many equilibrium conditions there are, and how many unknown bar tensions. If there are b bars, then there are b unknown tensions. Most structures are supported on foundations, and unknown reaction components are present. At a roller support of a plane framework (Figure 3.1(g)) one unknown reaction component is present, whereas at a simple support (Figure 3.1(f)) two components can be present. Denote the number of unknown reaction components by r. Altogether, then, there are $b+r$ unknowns.

In a plane framework, two equilibrium equations can be written for each joint. If there are j joints, that gives $2j$ equilibrium equations. It may be, however, that the external loads have themselves to obey certain equilibrium conditions: in Figure 3.19(i), for instance, the external loads must obey three conditions (that their vertical resultant be zero, that their horizontal resultant be zero, and that their resultant moment about any point be zero), since otherwise the framework cannot possibly be in equilibrium. If there are e such conditions, then only $2j-e$ of the $2j$ joint equations are independent. Accordingly, we expect a framework to be statically determinate if

$$b+r=2j-e, \tag{3.5.1}$$

statically indeterminate if

$$b+r>2j-e, \tag{3.5.2}$$

and a mechanism if

$$b+r<2j-e. \tag{3.5.3}$$

In Figure 3.18(i)

$$b=5, \quad r=0, \quad j=4, \quad \text{and} \quad e=3,$$

and so the framework is statically determinate. In Example 3.1.1 (Figure 3.6, page 29)

$b=4$,

$r=4$ (because there are two unknown reaction components at each of the two supports),

$j=4$ (counting the joints at the foundation), and

$e=0$.

This framework too is statically determinate. In the framework of Example 3.3.1 (Figure 3.15(a), page 39), $b=23$ and $j=13$. There are two reactions between the structure and its foundation, the vertical reactions at A and G, and so $r=2$. Whatever the external loads are, they cannot have any horizontal resultant, since otherwise the whole framework would roll sideways: that gives one condition on the loads, and so $e=1$. Accordingly

$$b+r=25=2j-e$$

and this framework is statically determinate.

These rules are not foolproof, and need to be used sensibly. Consider the framework shown in Figure 3.19(k). It has

$$b=9, r=3, j=6, e=0,$$

and so

$$b + r = 12 = 2j - e.$$

However, it is not statically determinate, but a mechanism. The right-hand panel can freely distort into a parallelogram, while the left-hand panel is in fact statically indeterminate. For the rules to apply, the bars must be properly arranged. A mathematical description of what proper arrangement means would be complex, and for this reason it is in case of doubt better to go back to the simpler idea of removing bars one by one until the framework becomes a mechanism.

In three-dimensional frameworks, there are as we have seen three equilibrium equations for each joint. The condition for statical determinacy becomes

$$b + r = 3j - e. \qquad (3.5.4)$$

Thus, the framework of Example 3.4.1 (Figure 3.17, page 41) is statically determinate and has $b = 9$, $r = 9$ (three reaction components for each of the three points of attachment to the foundation), $j = 6$, and $e = 0$. If, to give another example, a tetrahedral framework is made from six bars, and is not attached to any support, then $b = 6$, $j = 4$, $r = 0$, and $e = 6$ (because forces acting on the framework have to obey six equilibrium conditions for it to be in equilibrium), and it is therefore statically determinate.

3.6. Problems

1. Figure 3.20 shows a plane pin-jointed truss simply supported at A and resting on a roller support at C. Determine the bar tensions produced by
   (a) a vertical downward load P at joint E,
   (b) a horizontal load Q, acting to the right at joint E.

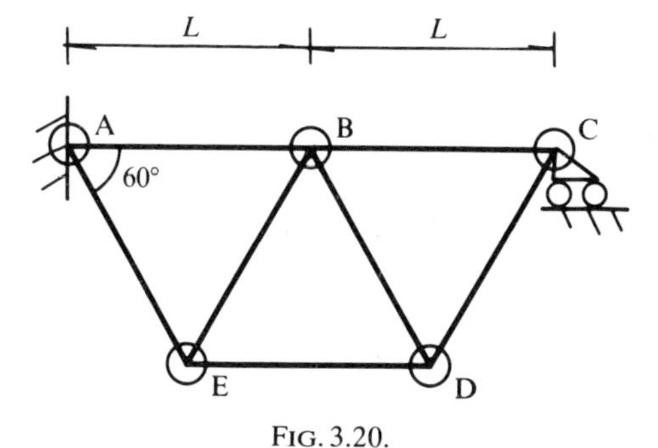

FIG. 3.20.

2. Figure 3.21 shows a plane pin-jointed framework, representing an idealization of a tower used to support hoses for unloading oil from tankers. Determine the tensions in the bars induced by a horizontal load of 100 kN at C.

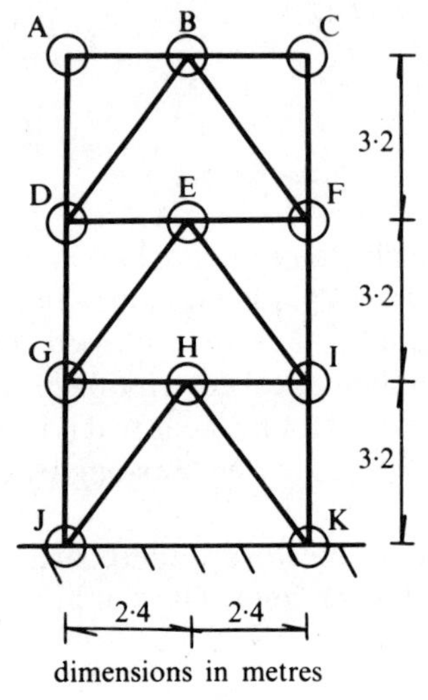

FIG. 3.21.

3. A simply supported plane bridge truss has a layout of the type shown in Figure 3.22. It has n panels, each a broad and b high, so that its span is na. It carries a unit vertical load at a point on the lower chord distant ia from the left support, where i is an integer less than n. Use the method of sections to derive a formula for the tensions induced in the bars of the jth panel, where $j \leq i$.

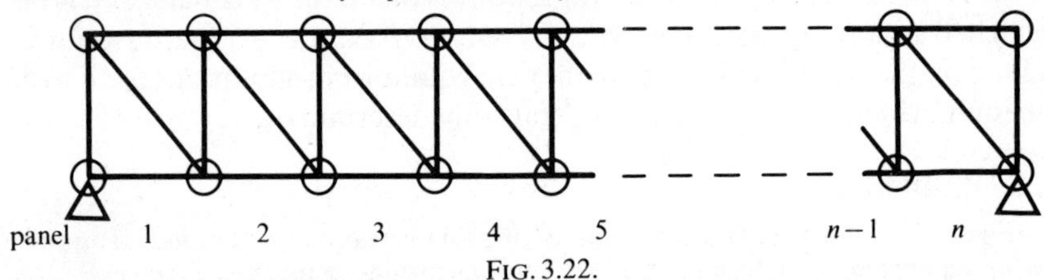

FIG. 3.22.

4. Figure 3.23 is an isometric view of a pin-jointed framework pinned to a rigid wall at A, C, and E. The coordinates of the joints are A (0, 0, 0), B (2, 0, 0), C (0, 0, 2), D (2, 3, 0), and E (0, 3, 0). The 3-axis is vertical. A vertical load of 10 kN acts at D. Calculate the tensions induced in the bars of the framework.

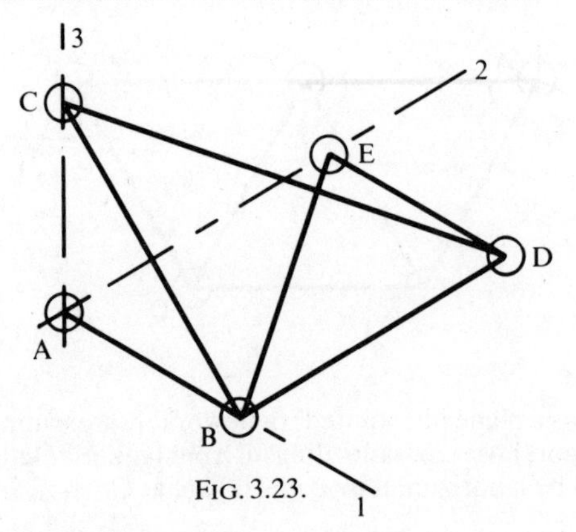

FIG. 3.23.

4. Internal forces in structures: beams and frames

4.1. Internal forces in plane structures

ONCE we leave pin-jointed frameworks and consider a more general class of skeletal structures, the internal forces need a more complex description. Imagine that a structure is in equilibrium under a number of external forces and moments (Figure 4.1(a)), and suppose that an imaginary sectioning

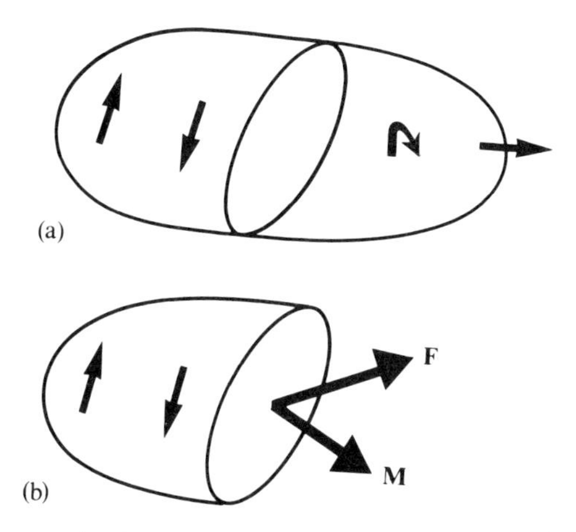

FIG. 4.1. (a) Structure under the equilibrium of external loads and moments. (b) Segment to the left of imaginary sectioning surface, in equilibrium under external loads and loads and moments across surface.

surface cuts the structure in two. Consider the segment of the structure to the left of the sectioning surface (Figure 4.1(b)). This segment will not generally be in equilibrium unless a resultant force is transmitted across the section. This force, exerted by one segment on the other, can be found in the usual way, since the vector sum of all the forces acting on the segment must be zero, this sum including both the forces acting across the imaginary section and those acting externally on the segment. In the same way, if we take moments about a point in the sectioning plane, we find that equilibrium also requires a resultant moment. The force and the moment are shown as vectors **F** and **M** in Figure 4.1(b): they will not generally have the same direction, or be oriented in any particular way with respect to the section.

The **F** and **M** vectors shown in Figure 4.1(b) represent the force and moment exerted on the left-hand segment by the right-hand segment. The force and moment exerted at the same section on the right-hand segment by the left-hand segment are equal and opposite.

Again we first consider plane structures, in which all the members lie in one plane and all the loads lie in the same plane. It follows that

(i) the force vector **F** across an imaginary cut lies in the plane of the structure, and

(ii) the moment **M** is perpendicular to the plane of the structure.

We shall be concerned with skeletal structures which can be represented by lines (not necessarily straight lines). An element of a plane skeletal structure is shown in Figure 4.2(a). In order to describe the internal forces at

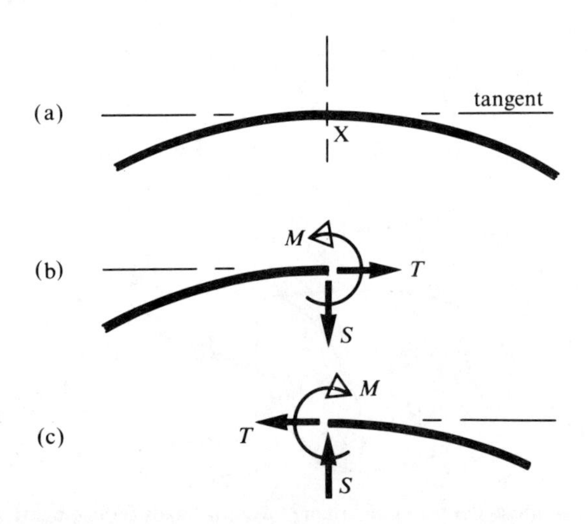

FIG. 4.2. (a) Element of plane skeletal structure. (b) Forces and moment of segment to the left of cut at X. (c) Forces and moments of segment to the right of cut at X. (d) Relation between sign convention and quantity describing position on segment.

a point X, imagine a cut at X. The resultant force exerted on the segment to the left of X is resolved into two components, one in line with the tangent to the element at X, the other perpendicular to it. The first component is called the tension, denoted T; it corresponds to the tension in a bar of a pin-jointed framework, which we met in Section 3.1. The second component, perpendicular to the tangent at X, is called the *shear force*, denoted S. The two force components are shown in Figure 4.2(b). The moment vector must be perpendicular to the plane of the structure, and is therefore completely described by its magnitude M, called the *bending moment*; it is positive if it acts counter-clockwise on the segment to the left of the cut. Figure 4.2(c) shows the forces and moment exerted on the segment to the right of the cut: they have the same magnitude, but are opposite in direction.

Because it refers to 'left' and 'right' sides of an imaginary cut, this sign convention is not sufficiently general to apply to all frames. To get a more general definition, we arbitrarily choose a direction along a skeletal structure member as the positive one. We can think of an arrow parallel to the member: the positive direction is the one in which the arrow points. This arrow does no more than define a direction, and has nothing to do with forces. An imaginary cut divides the member into two segments. If the bending moment is positive, a clockwise moment acts on the end of the segment towards which the direction arrow points, and an anti-clockwise moment acts on the end of the other segment (which the direction arrow points away from). If the bending moment is negative, an anti-clockwise moment acts on the end of the segment the direction arrow points towards, and a clockwise moment acts on the other segment. Comparison with the more restricted sign convention for beams, expressed in Figure 4.2, shows that the two definitions are consistent if the positive direction is from left to right.

It is often convenient to refer to points on elements by some variable which measures their position relative to some reference point, and it is then helpful consistently to define the positive direction as the one in which the variable increases. Examples of this approach will be found in Sections 4.2 and 7.1.

Example 4.1.1. A simply supported beam 3 m long carries a concentrated load of 4·5 kN at 2 m from one end (Figure 4.3(a)). What are the tension, shear force, and bending moment induced in it?

The term 'simply supported' means that the supports can exert vertical and horizontal forces, but not moments.

The quantities we are concerned with can be expected to vary from point to point along the beam, and so we need a notation to locate points on the beam. The distance from the left-hand support will be called x, so that

$x = 0$ at the left-hand support,

$x = 2$ at the load,

$x = 3$ at the right-hand support.

Support reactions are calculated in the usual way, as in Section 2.3, and are 1·5 kN at the left-hand support and 3 kN at the right-hand support.

Consider first of all a point X_1 to the left of the load. Imagine a cut at X_1, and draw the segment of the beam to the left of X_1. Figure 4.3(b) shows the forces and moments acting on the segment. At the left-hand end there is a vertical reaction of 1·5 kN and no moment. At the right-hand end of the segment, at the cut, there are a tension T, a shear force S, and a bending moment M. The signs shown are consistent with Figure 4.2. The segment to the left of X_1 is in equilibrium. Writing down the three equilibrium

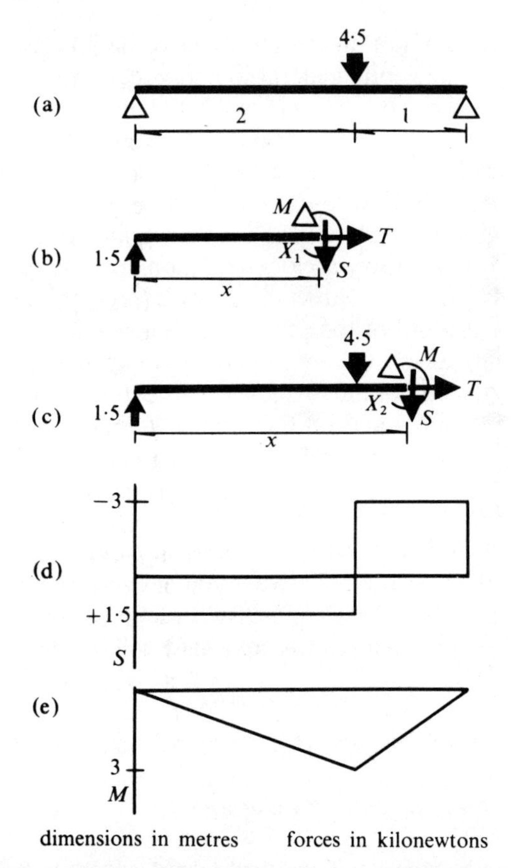

FIG. 4.3. Simply supported beam under concentrated load.

conditions

$$T = 0, \tag{4.1.1}$$

since the horizontal force resultant on the segment is zero,

$$1{\cdot}5 - S = 0, \tag{4.1.2}$$

since the vertical force resultant on the segment is zero, and

$$1{\cdot}5x - M = 0, \tag{4.1.3}$$

taking moments about X_1. Accordingly

$$\left.\begin{aligned} S &= +1{\cdot}5 \text{ kN} \\ T &= 0 \\ M &= 1{\cdot}5x \text{ kN m} \end{aligned}\right\} \text{for } x < 2 \text{ m.} \tag{4.1.4}$$

These expressions only apply to the left of the load. Imagine a new cut at a point X_2, to the right of the load, and consider the forces on the segment to the left of the cut (Figure 4.3(c)). Since the segment is in equilibrium, we can again apply the three conditions, and then

$$\left.\begin{aligned}&\text{(vertical equilibrium)} & S&=-3\text{ kN}\\ &\text{(horizontal equilibrium)} & T&=0\\ &\text{(moments about } X_2) & M&=1{\cdot}5x-4{\cdot}5(x-2)\end{aligned}\right\}\text{ for } x>2\text{ m.}\quad(4.1.5)$$

We know that at $x=3$ there is a simple support, at which the bending moment is zero and the shear force must balance the 3 kN support reaction. This gives us a check on eqns (4.1.5), which ought to agree with these conditions, and do so.

It was convenient to take moments about the cut X_1 or X_2 to find M, because then S and T do not enter the moment equation. If moments are taken about a different point, the same results appear when S and T are put into the resulting equation. The decision to consider equilibrium of the segment to the left of the cut was an arbitrary one: the right-hand segment would have done equally well, and have given the same results, as you can check for yourself.

The distributions of shear force and bending moment can be plotted graphically. This is done in Figure 4.3(d) and (e). By a convention of beam analysis, shear force and bending moment are plotted positive downward.

Example 4.1.2. A cantilever of length L carries a uniformly distributed downward load of intensity w per unit length (Figure 4.4(a)). What are the shear force and bending moment at a distance x from the built-in end?

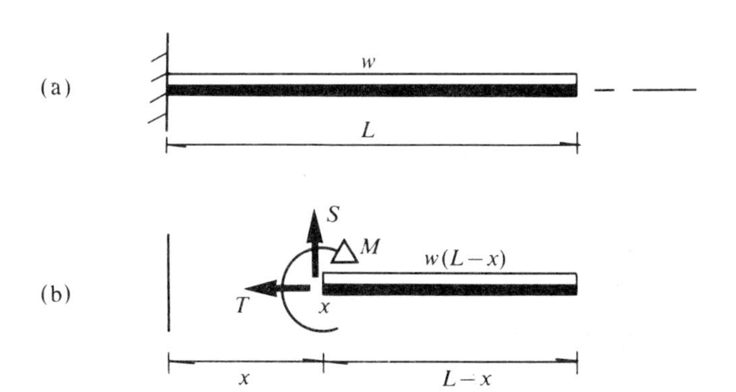

FIG. 4.4. Cantilever under distributed load.

Imagine a cut X, at a distance x from the built-in end. The segment to the right of the cut is shown in Figure 4.4(b). It is easier to consider this segment than the one to the left, because it makes it possible for us to avoid working out the reactions at the left-hand end. The length of this segment is $(L-x)$, and so the total load on it is $w(L-x)$, acting downwards. Since the segment is in equilibrium

$$\begin{aligned} S &= w(L-x) \\ T &= 0. \end{aligned} \tag{4.1.6}$$

The segment load $w(L-x)$ is uniformly distributed. When we consider the equilibrium of the segment, by taking moments about X, the moment of the segment load about X is the same as the moment of a concentrated force $w(L-x)$ acting half-way along the segment, at $\frac{1}{2}(L-x)$ from X. Taking moments about X, then

$$M = -\underbrace{\{w(L-x)\}}_{\text{load}}\underbrace{\{\tfrac{1}{2}(L-x)\}}_{\text{lever arm}} = -\tfrac{1}{2}w(L-x)^2. \tag{4.1.7}$$

At the free end x is L, no external load or moment act on the cantilever, and S and M must both be zero. The expressions we have derived are consistent with this.

Example 4.1.3. A beam resting on two simple supports carries the loading shown in Figure 4.5. Calculate the shear force and bending moment distributions along its length.

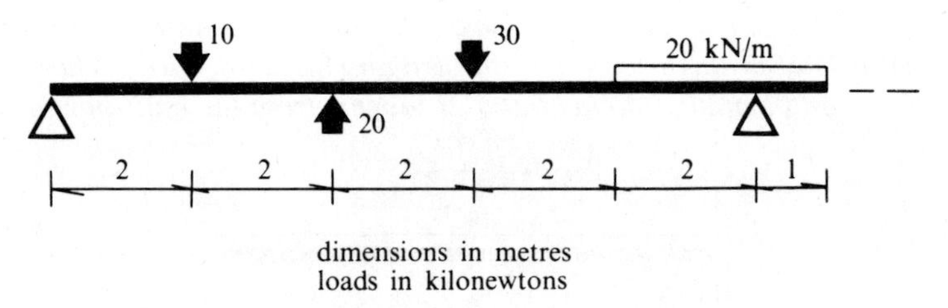

FIG. 4.5. Simply supported beam carrying several loads.

The end reactions are 11 kN at the left and 69 kN at the right. Define position on the beam by distance x from the left-hand end. The natural way to proceed is to follow the technique used in Example 4.1.1. First we imagine a cut between the left-hand end and the 10 kN load, and use the conditions for equilibrium of the segment to the left of the cut, which gives

$$\left.\begin{aligned} S &= 11 \\ M &= 11x \end{aligned}\right\} \quad x<2. \tag{4.1.8}$$

Next, imagine a cut between the 10 kN load and the 20 kN load, and use equilibrium conditions for the segment to the left of that cut, which gives

$$\left.\begin{aligned} S &= +11-10=1 \\ M &= 11x-10(x-2). \end{aligned}\right\} \quad 2<x<4. \tag{4.1.9}$$

Next

$$\left.\begin{aligned} S &= +11-10+20=21 \\ M &= 11x-10(x-2)+20(x-4), \end{aligned}\right\} \quad 4<x<6 \tag{4.1.10}$$

and so on. Obviously this becomes tedious if there are a lot of loads, and one wonders if it might not be possible to write down single expressions for S and M which would apply to the whole beam. Looking back to the expressions above, we see that eqns (4.1.9) for the second zone $2<x<4$ could have been got by adding to eqns (4.1.8) additional terms corresponding to the 10 kN load which has been 'passed' at $x=2$. Alternatively, eqns (4.1.8) can be got from (4.1.9) by striking out the terms corresponding to the 10 kN load, which only appear if $x>2$. Similarly, eqns (4.1.9) can be got from (4.1.10) by striking out the terms corresponding to the 20 kN load, which only appear if $x>4$.

Macauley (1919) developed a systematic notation to keep track of which terms to leave out. It needs only a distinctive kind of bracket, which we shall write [ ]. In equations, terms containing this special bracket will only count if the quantity within the bracket is positive. If the quantity within the bracket is negative, the term is ignored. Thus

$$[x] \text{ will mean } \begin{cases} x, \text{ if } x \text{ is positive,} \\ 0, \text{ if } x \text{ is negative,} \end{cases}$$

$$[x]^2 \text{ will mean } \begin{cases} x^2, \text{ if } x \text{ is positive,} \\ 0, \text{ if } x \text{ is negative,} \end{cases}$$

$$[x+3] \text{ will mean } \begin{cases} x+3, \text{ if } x>-3, \\ 0, \text{ if } x<-3, \end{cases}$$

but

$$[x]+3 \text{ will mean } \begin{cases} x+3, \text{ if } x>0, \\ 3, \text{ if } x<0. \end{cases}$$

Note that $[x+3]$ and $[x]+3$ are not the same. Since any quantity to the power zero is 1, we can also say that

$$[x]^0 \text{ will mean } \begin{cases} 1, \text{ if } x>0, \\ 0, \text{ if } x<0. \end{cases}$$

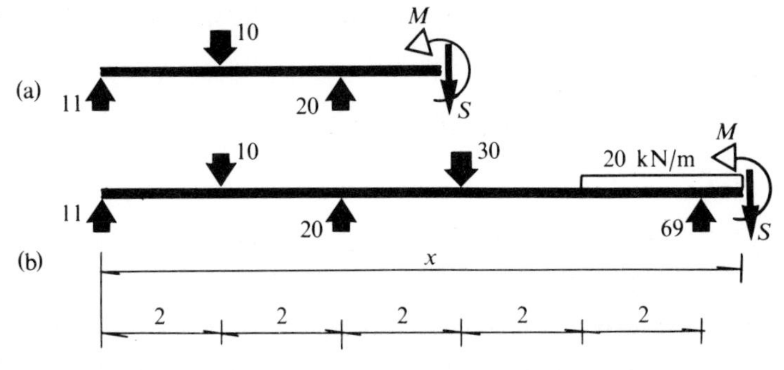

FIG. 4.6. Simply supported beam of Example 4.1.3. (a) Forces and moment on segment to the left of an imaginary cut between the 20 kN and 30 kN loads. (b) Forces and moment on segment to the left of an imaginary cut between the right-hand support and the end.

Suppose, then, that we return to the example, and imagine a cut between the 20 kN and 30 kN loads. The segment to the left of this cut is shown in Figure 4.6(a). Taking moments about the cut, we can write

$$M = 11[x] - 10[x-2] + 20[x-4]. \tag{4.1.11}$$

This will apply anywhere to the left of the cut, provided we follow the notation described above, and leave out terms if a negative quantity appears in the Macauley bracket. Similarly

$$S = 11[x]^0 - 10[x-2]^0 + 20[x-4]^0. \tag{4.1.12}$$

To derive an expression valid for the whole beam, take a cut to the right of all the concentrated forces, and consider equilibrium of a segment to the left of that cut (Figure 4.6(b)). Then

$$S = 11[x]^0 - 10[x-2]^0 + 20[x-4]^0 - 30[x-6]^0 - 20[x-8] + 69[x-10]^0 \tag{4.1.13}$$

$$M = 11[x] - 10[x-2] + 20[x-4] - 30[x-6] - 20[x-8]^2 + 69[x-10]. \tag{4.1.14}$$

This notation will be useful later when we come to study beam deflections in Chapter 5.

Example 4.1.4. A semi-circular arch of radius R is built-in at both ends (Figure 4.7(a)). At the right-hand abutment the horizontal reaction is H_B, the vertical reaction V_B, and the moment M_B. Find the tension, shear force, and moment at a general point on the arch.

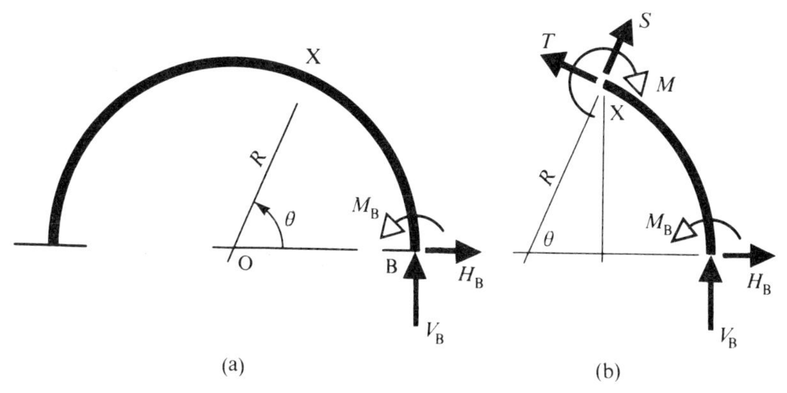

FIG. 4.7. (a) Semi-circular arch. (b) Forces and moments on segment of arch.

The shape of the arch makes it awkward to use x, y coordinates to locate points. It is much simpler to locate points by an angle θ subtended at the centre, measured counter-clockwise from OB. Imagine a cut at X, located by θ, and consider the equilibrium of the segment between X and B (Figure 4.7(b)). It is in equilibrium under the reaction forces H_B and V_B at B, the reaction moment M_B, and the tension T, shear force S, and moment M at X.

It is simplest to equate force resultants in the local tangential and normal directions to the arch at X, because to do so separates T and S. Since the resultant force on the segment in the tangential direction at X must be zero,

$$T - H_B \sin\theta + V_B \cos\theta = 0$$

$$T = H_B \sin\theta - V_B \cos\theta. \qquad (4.1.15)$$

Since the resultant force on the segment in the normal direction at X must be zero,

$$S + H_B \cos\theta + V_B \sin\theta = 0$$

$$S = -H_B \cos\theta - V_B \sin\theta. \qquad (4.1.16)$$

Taking moments about X,

$$M - H_B(R \sin\theta) - V_B R(1 - \cos\theta) - M_B = 0$$

$$M = H_B R \sin\theta + V_B R(1 - \cos\theta) + M_B. \qquad (4.1.17)$$

4.2. Relation between shear force and bending moment

Consider a straight beam which carries a distributed load of intensity $q(x)$, positive upward and not necessarily constant. Both the bending moment M and the shear force S will vary along the beam. As before, denote distance along the beam by x, increasing to the right, measured from an arbitrary

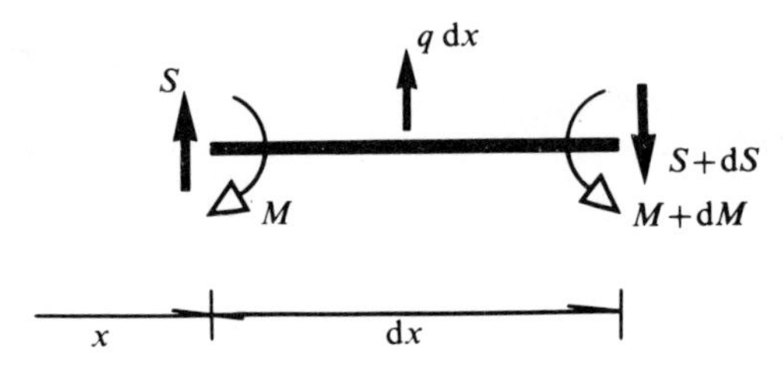

FIG. 4.8. Element of a straight beam.

origin. A short element of the beam is shown in Figure 4.8: its left-hand end is at x, and it is $\mathrm{d}x$ long, so that the load it carries is $q\mathrm{d}x$. The forces on it are S at the left-hand end, $S+\mathrm{d}S$ at the right-hand end (because S is varying); the signs in the Figure are consistent with the definition in Figure 4.2. The bending moments are M at the left, $M+\mathrm{d}M$ at the right (because M is varying too), and again the signs are consistent. Now consider the equilibrium of the element. It gives

resolving
vertically $\quad S+q\,\mathrm{d}x-(S+\mathrm{d}S)=0 \qquad (4.2.1)$

taking
moments $\quad M+(S+\mathrm{d}S)\,\mathrm{d}x-(q\,\mathrm{d}x)(\tfrac{1}{2}\,\mathrm{d}x)-(M+\mathrm{d}M)=0. \qquad (4.2.2)$

In the limit, as $\mathrm{d}x$ tends to zero, these equations become

$$\frac{\mathrm{d}S}{\mathrm{d}x}=q \tag{4.2.3}$$

$$\frac{\mathrm{d}M}{\mathrm{d}x}=S. \tag{4.2.4}$$

These relations always apply. They can of course be derived in a mathematically more rigorous way. If you look back to the expressions for shear force and bending moment derived in various earlier examples, for instance eqns (4.1.6) and (4.1.7) for Example 4.1.2, you will see that they apply there. We shall use them again in Chapter 7.

4.3. Three-dimensional structures

In three-dimensional skeletal structures the resultant force **F** and moment **M** across a section of a member can have any orientation with respect to each other and to the member (Figure 4.2). It is useful to generalize the concepts of tension, shear force, and bending moment, stress resultants so far defined only for plane structures loaded in the same plane. A generalization is necessary if we are to be able properly to relate the external loading to the internal forces, and also because the local deformation of a structural element is related to the local values of the stress resultants.

Again consider a cut through a structural element. Construct a tangent to the element at the cut, and define local reference axes 1′, 2′, and 3′, in such a way that the 1′-axis coincides with the tangent and the 2′- and 3′-axes are perpendicular to the tangent and to each other (Figure 4.9). The force

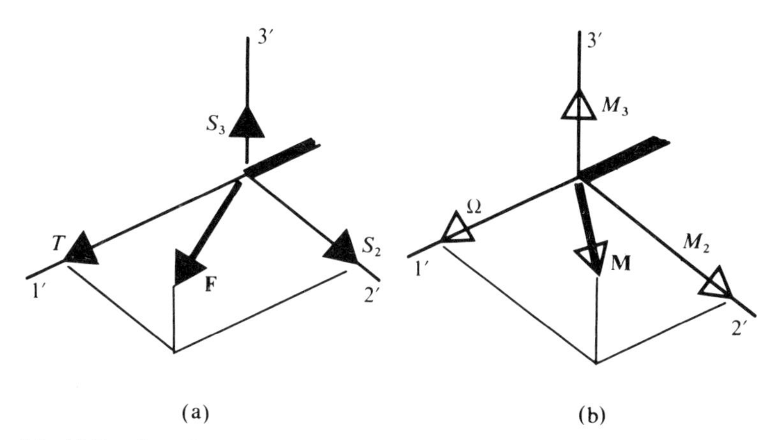

FIG. 4.9. (a) Resultant force across a cut in a three-dimensional skeletal structure, and its components. (b) Resultant moment across a cut in a three-dimensional skeletal structure, and its components.

resultant **F** across the cut can be resolved into components in the three coordinate directions, as in Figure 4.9(a), so that

the component of **F** in the 1′-direction is the tension T,
the component of **F** in the 2′-direction is the first shear force S_2,
the component of **F** in the 3′-direction is the second shear force S_3.

The moment **M** can be divided into three components in the same way (Figure 4.9(b)), so that

the component of **M** in the 1′-direction is the torque Ω,
the component of **M** in the 2′-direction is the first bending moment M_2,
the component of **M** in the 3′-direction is the second bending moment M_3.

These are consistent with the earlier definitions for plane structures. The definition of torque is consistent with the elementary definition for a straight element subjected to pure torsion.

The definition of the 2′- and 3′-directions does not fix their orientation normal to the local tangent. In structural stress analysis, which is beyond the scope of this book, it is often useful to relate these axes to the cross-section of the member concerned. If, for instance, it is an I-section, which has two perpendicular axes of symmetry, it would be natural to make the 2′- and 3′-axes coincide with them.

Example 4.3.1. In the pipe considered in Example 2.3.5 (Figure 2.12, page 19), what are the tension, shear forces, torque, and bending moments at the midpoint of BC?

Imagine a cut at X, the midpoint of BC, and consider the equilibrium of ABX. It is subjected to

(i) a force $4{\cdot}12\mathbf{i}+5\mathbf{j}+5{\cdot}09\mathbf{k}$ at A (from the results of Example 2.3.5)
(ii) a torque $1{\cdot}08\mathbf{j}$ at A,
(iii) the weight $-1{\cdot}2\mathbf{k}$ of AB, which acts at $0\mathbf{i}+1\mathbf{j}+0\mathbf{k}$,
(iv) the weight $-0{\cdot}6\mathbf{k}$ of BX, which acts at $0{\cdot}5\mathbf{i}+2\mathbf{j}+0\mathbf{k}$,
(v) an unknown force $\mathbf{F}$ at X,
(vi) an unknown moment $\mathbf{M}$ at X.

Since the vector sum of the forces on ABX is zero:

$$0=(4{\cdot}12\mathbf{i}+5\mathbf{j}+5{\cdot}09\mathbf{k})-1{\cdot}8\mathbf{k}+\mathbf{F}$$

$$\mathbf{F}=-4{\cdot}12\mathbf{i}-5\mathbf{j}-3{\cdot}29\mathbf{k} \qquad (4.3.1)$$

The tension is the component of this force in the local axial direction of the pipe, which is the 1-direction, and is therefore $-4{\cdot}12$ kN. The shear forces in the 2- and 3-directions are -5 kN and $-3{\cdot}29$ kN respectively.

Since the resultant moment about X is zero

$$0=\{(4{\cdot}12\mathbf{i}+5\mathbf{j}+5{\cdot}09\mathbf{k})\times(-1\mathbf{i}-2\mathbf{j}+0\mathbf{k})\}+1{\cdot}08\mathbf{j}$$
$$+\{(-1\mathbf{i}-2\mathbf{j})\times(-1{\cdot}2\mathbf{k})\}+\{(-0{\cdot}5\mathbf{i})\times(-0{\cdot}6\mathbf{k})\}+\mathbf{M}, \qquad (4.3.2)$$

whence

$$\mathbf{M}=-12{\cdot}58\mathbf{i}+5{\cdot}51\mathbf{j}+3{\cdot}24\mathbf{k}. \qquad (4.3.3)$$

The torque is therefore $-12{\cdot}58$ kN m, and the bending moments in the 2- and 3-directions are $5{\cdot}51$ kN m and $3{\cdot}24$ kN m respectively.

4.4. Problems

1. A simply supported beam carries the loading shown in Figure 4.10. Find
(a) the support reactions,
(b) the bending moment at the left-hand support,
(c) a general expression for the shear force and bending moment at a distance x from the left-hand end.

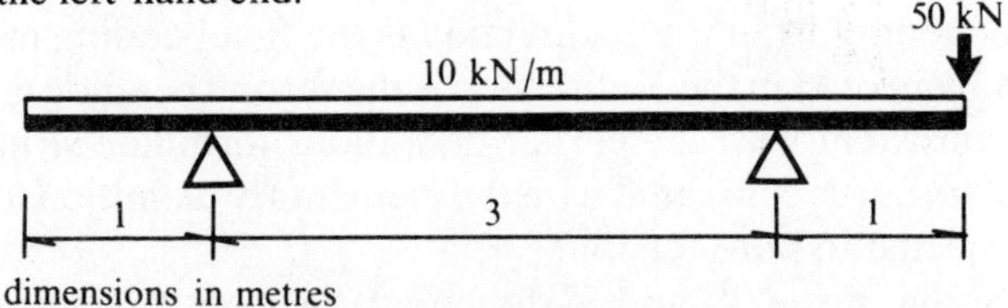

FIG. 4.10.

2. A beam of length L is simply supported at both ends, and carries a load which increases linearly from 0 at the left support to q at the right. Calculate
(a) the support reactions,
(b) the shear force and bending moment at a point at a distance x from the left-hand support,
(c) the location and magnitude at the maximum bending moment.

3. A parabolic arch has a height h and span L. Its ends are pinned to rigid abutments, which are at the same level, and there is a third hinge at the centre of the arch. The left-hand half of the arch carries a load $qL/2$, which is uniformly distributed *horizontally* (rather than along the length of the arch rib); the right-hand half of the span is unloaded. Calculate
   (a) the vertical and horizontal reactions at the abutments,
   (b) the maximum bending moment in the left-hand segment,
   (c) the maximum bending moment in the right-hand segment.

4. Figure 4.11 shows a plane portal frame, with stanchion feet built-in to the foundation. It carries a load of 200 kN uniformly distributed along the beam. The reactions at the right-hand support are found to have the values shown. Find
   (i) the reactions at the left-hand support,
   (ii) the tension, bending moment, and shear force at the centre of the beam,
   (iii) the tension, bending moment, and shear force in the right-hand stanchion at a distance y above the support,
   (iv) the tension, bending moment, and shear force in the beam at a distance x to the right of the top left-hand corner.

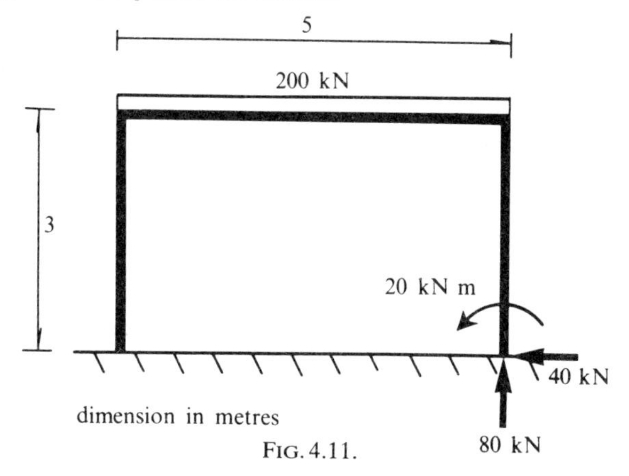

FIG. 4.11.

5. Figure 4.12 is a plan view of a U-shaped beam which rests at A, C, and E on supports which can only exert vertical reactions. Its own weight is 2 kN/m, uniformly distributed, and it carries a vertical load of 10 kN at B. Calculate the shear force, bending moment, and torque at a point distance x along BC from B.

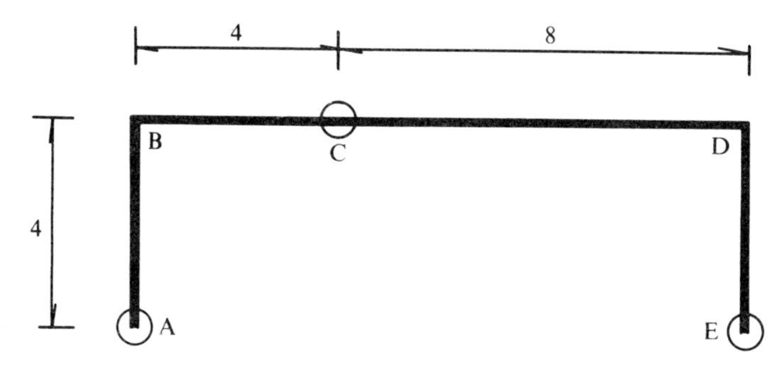

FIG. 4.12.

5. Deformations of Structures

5.1. Introduction

An engineer who designs a structure is not only concerned with the forces within it but also with how far it distorts under load. Sometimes there are direct constraints on allowable distortions. A factory building will clearly be unsatisfactory if its floors and walls deflect too far under the weight of people and equipment, even if they do so without collapsing, for the people will feel unsafe, and some of the equipment, such as lifts, will no longer work properly. Excessive deflections of an aircraft wing will alter its aerodynamic characteristics, and can lead to an instability in which the whole wing flutters. A 1 mm deflection of a radio telescope aerial under wind loading will make it unserviceable, because any alteration of the relative position of different sectors of the aerial confuses the incoming signal. Sometimes, again, an engineer is interested in the deformations of a structure less for their own sake than because of their influence on the forces set up within it. This is the case with the statically indeterminate structures we met in the previous chapter, where it turns out that the forces can only be found after a study of the distortion that the loaded structure can undergo.

In this chapter we first develop a notation to describe deflections, and then examine the relative movements between one part of a structure and another. It happens in a few simple cases that the deflections can be found very easily, without any further theory. In more complicated cases we need the principle of virtual work, and that is developed in the next chapter.

5.2. Displacements

Imagine a structure to occupy an initial configuration represented by the dashed outline in Figure 5.1. Two particles of the structure are then in the positions located by A and B in the Figure, located by position vectors **a** and

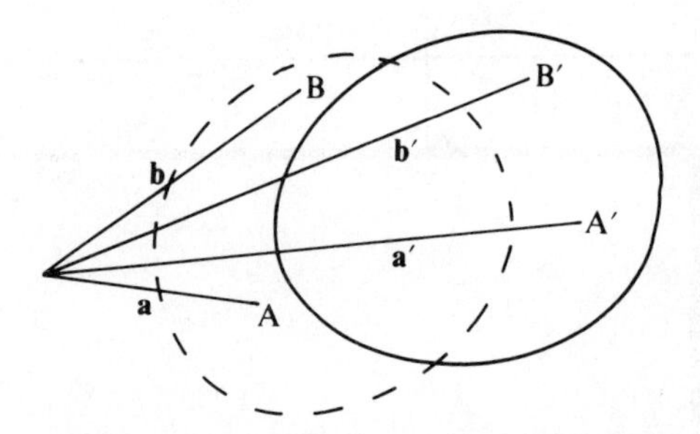

FIG. 5.1. Deformation of a structure, and definition of displacement.

b. The structure then moves and distorts in some way, and takes up a new configuration represented by the solid outline. The particle that was in position A moves to A′, located by **a**′, and the particle that was in position B moves to B′, located by **b**′. The vector difference **a**′ − **a**, represented by AA′ in the Figure, is called the displacement of A. Similarly, the vector difference **b**′ − **b**, represented by BB′, is the displacement of B. Displacement is a vector quantity: it will be denoted by **u**, and when we need components they will be denoted by u, v, and w, in the 1-, 2-, and 3-directions.

A good way to represent displacement is to draw a displacement diagram, a diagram in which the body which is deforming does not appear but the motions of individual particles do. In such a diagram, Figure 5.2, o represents all points which do not move at all, so that the position they occupy in

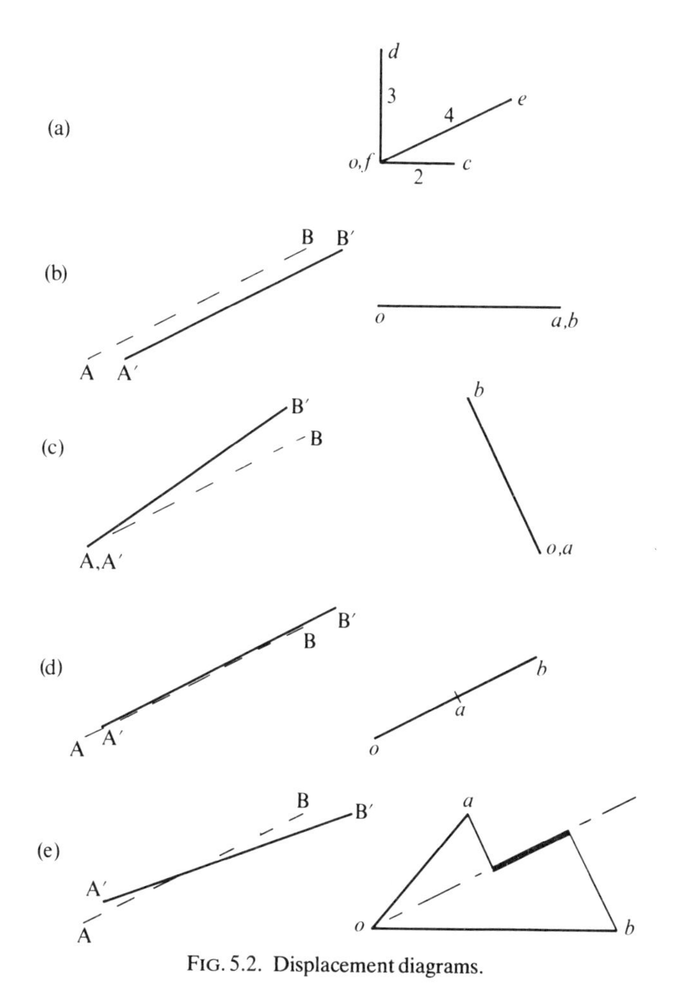

FIG. 5.2. Displacement diagrams.

the deformed configuration is the same as that in the initial reference configuration. Points indicate displacements, and are labelled by lower-case letters, so that the point corresponding to the motion of A is labelled *a*, and so on. Figure 5.2(a) is an example: it indicates that C has moved 2 mm to the right, that D has moved 3 mm upwards, that E has moved upwards and to the right through 4 mm at 25° to the horizontal, and that F has not moved. These are movements from the initial position, but relative movements are given correctly by the diagram, so that the movement of C relative to D is represented by *dc*.

Imagine a structure consisting of a single straight bar AB. Figure 5.2(b–e) shows some particularly simple kinds of motion of the bar, and the corresponding displacement diagrams. The left-hand side of the Figure shows the bar in its initial (dashed line) and final (solid line) configurations, for the different motions, and the right-hand side shows the corresponding displacement diagram, drawn to a larger scale so that each displacement is five times larger than it appears in the diagram showing the bar itself. In Figure 5.2(b) the bar moves sideways without rotating or stretching, so that both ends of the bar have identical displacements. In Figure 5.2(c) end A does not move at all, so that in the displacement diagram *o* and *a* coincide, but the whole bar rotates about A. Even though the bar remains the same length, there is relative motion between A and B. As long as the rotation is small, the relative displacement *ab* is perpendicular to AB and equal to the bar length multiplied by the angle it rotates through. The sense in which the word 'small' is used here will be discussed a little further on. In Figure 5.2(d) there is no rotation, but although the direction of the rod after the motion is the same as it was before, the distance between A and B has increased, and so the bar is longer than it was. The amount by which the length of the bar has increased is called the *elongation*; in Figure 5.2(d) the elongation is represented by *ab*.

We can see in Figure 5.2(c) that the fact that there is relative motion between two points does not necessarily mean that the distance between them has increased: there may just have been rotation. When structures deform their members generally both rotate and stretch. The stretches are related to tensions in the bars, and to other effects like thermal expansion. It will clearly be necessary to distinguish the component of relative displacement associated with stretching from that associated with rotation. Looking at Figures 5.2(c) and 5.2(d), we can see that stretching without rotation produces a relative displacement aligned along the bar itself, and that rotation without stretching produces a relative displacement perpendicular to the bar (as long as the rotation is small).

Figure 5.2(e) shows a general motion, which involves both rotation and stretching. The corresponding elongation of the bar is the component of the relative displacement *ab*, resolved along the direction of the bar itself: this elongation is indicated by the heavy line. The other component, perpendicular to the bar, is associated with rotation, and is equal to the angle through which the bar has rotated multiplied by its length.

If a large rotation had occurred, the relative displacement between A and B in Figure 5.2(c) would not have been perpendicular to the bar, as you can see by thinking of what happens for a 90° rotation. When there are large rotations, the problem of separating the components due to rotation and stretching becomes more complex. In almost all real structures, however, any displacements that occur are small by comparison with the lengths of the members, and we can assume that rotations are small enough for the simplifications described above to be reasonable. Under the same conditions, we can avoid having to make distinctions between the original directions of bars and their directions after deformation, and between their original lengths and their deformed lengths. In most structures, these assumptions are quite reasonable, and from now on we shall tacitly assume them to hold, but in the back of our minds we need to remember that they will not apply to very flexible structures, such as balloons, or air-supported roofs, or cables mooring ships.

In dealing with frameworks, we denote the elongation of a bar IJ by e_{ij}. It is positive if it is tensile (so that the bar lengthens), negative if it is compressive (so that the bar shortens).

Example 5.2.1. Figure 5.3(a) shows a simple framework. It is loaded and its temperature changes. Measurement shows that as a result

A does not move,
B moves 5 mm downward, and 1 mm to the right,
C moves 1 mm downward, and 3·5 mm to the left,
D moves 2·2 mm downward, and 1 mm to the left.
What are the elongation and rotation of DC?

The displacement diagram is Figure 5.3(b); it can be constructed at once by plotting the given displacements from the reference starting point *o*. The relative displacement of D and C is *dc*, which in Figure 5.3(c) is resolved into two components, one along the direction of the bar DC and the other perpendicular to it. The first component is the change in length of DC, whose magnitude is, by trigonometry from the diagram,

$$(3{\cdot}5-1)\cos\theta+1{\cdot}2\sin\theta=2{\cdot}5\times0{\cdot}8+1{\cdot}2\times0{\cdot}6=2{\cdot}72\text{ mm}.$$

Has the bar got longer or shorter? If we look back to Figure 5.2(d), which referred to an elongation, the relative displacement of *b* to *a* was in the same direction as the relative physical positions of B and A. Point B was to the right of and up from A, and the relative displacement was in the same direction, so B had moved further in that direction, and the distance between

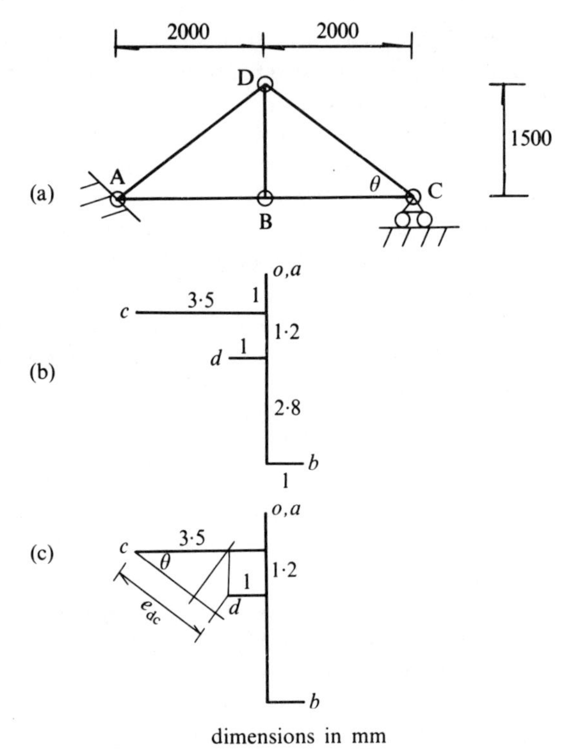

FIG. 5.3. Displacement diagram for Example 5.2.1.

B and A had increased. Returning to Figure 5.3(c), the component of *dc* in the bar direction is opposite to DC; D has moved down and to the right with respect to C, and so has moved closer to C. The bar has shortened, and therefore

$$e_{\mathrm{dc}} = -2{\cdot}72 \text{ mm}.$$

The component of *dc* perpendicular to DC is

$$(3{\cdot}5-1)\sin\theta - 1{\cdot}2\cos\theta = 2{\cdot}5\times 0{\cdot}6 - 1{\cdot}2\times 0{\cdot}8 = 0{\cdot}54\text{mm}.$$

The length of DC is 2500 mm, and so the rotation of DC is

$$0{\cdot}54/2500 = 2{\cdot}16\times 10^{-4} \text{ radians}$$

$$= 45 \text{ seconds of arc, clockwise.}$$

5.3. Deflections in pin-jointed frameworks

Figure 5.4(a) shows the framework that was analysed in Example 3.1.1, made from four bars hinged together and fixed to a rigid and immovable

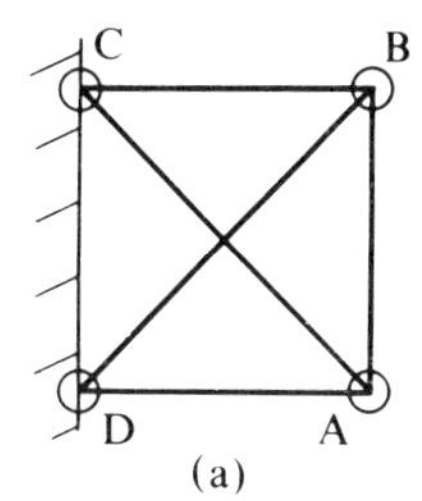

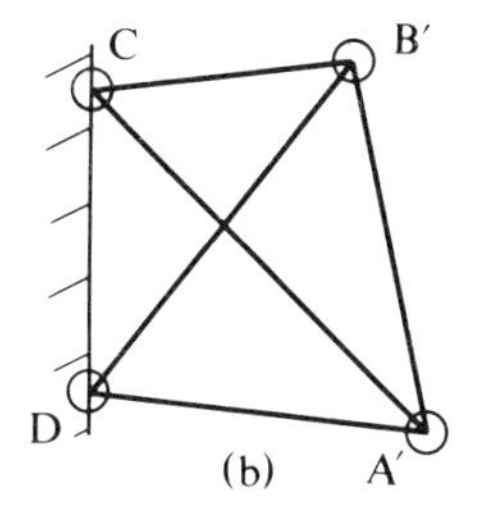

FIG. 5.4. Pin-jointed framework.

foundation at C and D. Imagine the lengths of the bars to alter slightly, by known amounts. The joints then move, so as to keep the distances between them equal to the lengths of the corresponding bars, and the framework takes up a new configuration, shown in Figure 5.4(b) with the displacements exaggerated. The problem of finding out how far the joints move is clearly a geometric one. It can be solved in several ways, for instance

(i) graphically, by drawing the undeformed structure with the original bar lengths, drawing the deformed structure with the altered bar lengths, and measuring the differences in the joint positions, or
(ii) analytically, by using trigonometry, or
(iii) by extending the idea of a displacement diagram, explained in Section 5.2 for a single bar, to represent the displacements of several joints.

Method (i) is hopelessly inaccurate, unless the elongations are very large indeed. Method (ii) is cumbersome and laborious, as you can confirm for yourself by trying one of the examples below. Method (iii) is useful, but becomes confusing for large structures. Figure 5.5 shows how the displacement diagram is constructed. Joints C and D are known not to move, because they are fixed to a foundation, and therefore they are represented by points c and d coincident with o. Bar CB is horizontal, and elongates by e_{cb}. It follows that the horizontal component of the displacement of B relative to C is e_{cb}, and that the point b in the displacement diagram lies on a vertical line at a distance e_{cb} to the right of c. As yet, we do not know where on the line b lies, but it must lie somewhere on the line if the horizontal component is to be correct. Similarly, bar BD is at 45° to the horizontal, and its elongation is e_{bd}. Point b therefore lies on a second line, distance e_{bd} from d in the direction of DB, at 45° to the horizontal. Once again, this does not tell us where b is on the line, but only that it must lie on it somewhere.

We now have two lines, and know that b lies on both of them. It must be at their intersection (Figure 5.5(a)), and so now the displacement of B is known. Notice that we have not needed to know the elongations of AB or AD, and if we look at Figure 5.4 we can see why that is: the lengths of bars BC and BD fix the position of B, whatever happens to the other bars.

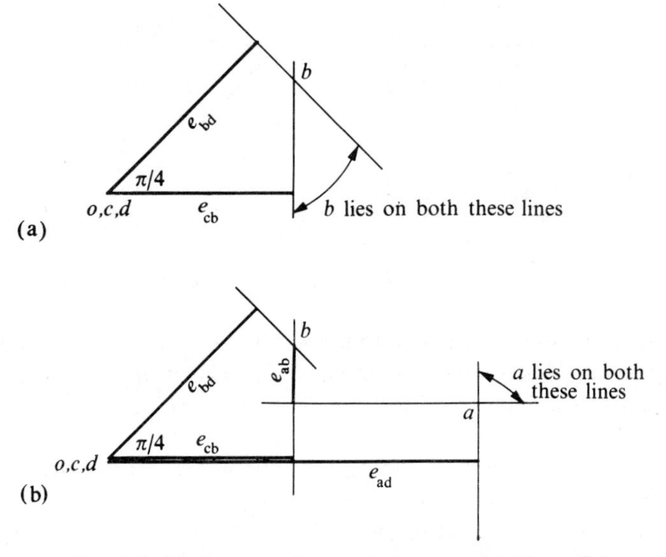

FIG. 5.5. Displacement diagram for framework in Figure 5.4.

The vertical component of the motion of A relative to B is e_{ab}. Now that b is located, we can draw a horizontal line e_{ab} below b, and know that a lies on it. Finally, the horizontal component of the motion of A relative to D is e_{ad}, and so a must lie on a vertical line e_{ad} to the right of d. Once again, there are two lines on which the point a we are looking for must lie, and their intersection locates it (Figure 5.5(b)).

If this construction had been carried out graphically, for particular values of the elongations, the displacements of B and C could simply have been measured off the diagram. If the diagram is a simple one, the displacement can also be found from the trigonometry of the diagram. From Figure 5.5(a), the vertical and horizontal displacements of B, written v_B and u_B, are

$$v_B = e_{bd}\sqrt{2} - e_{bc}$$

$$u_B = e_{bc},$$

and the displacements of A are, from Figure 5.5(b),

$$v_A = e_{bd}\sqrt{2} - e_{bc} - e_{ab}$$

$$u_A = e_{ad}.$$

These are purely geometric results, and depend only on the elongations of the bars and how they are connected together. It does not matter how the elongations are caused. They would still be correct if there were an additional bar connecting A to C.

The displacement diagram method is extensively developed in many textbooks of structural mechanics, and they explain how to overcome the extra difficulties that occur in more complicated structures, in which it is not easy to start the construction because the fixed points are at opposite ends of the framework. In most cases, however, it is easier to use the virtual work method described in the next chapter, and that has the additional advantage that it can be used for three-dimensional frameworks.

Example 5.3.1. The structure of Example 3.1.1 is made from uniform bars. The elongation of each bar is related to its tension by Hooke's law, so that

$$\textit{elongation} = \frac{(\textit{tension})(\textit{bar length})}{E(\textit{cross-sectional area})},$$

where E is a property of the material (Young's modulus). The cross-sectional area of bars BC, BA, and AB is 500 mm^2*, and that of BD is 1100 mm*^{2}*, and all the bars are steel, for which E is 210 kN/mm*^{2}*. The framework is loaded by a force of 100 kN, inclined at 30° to the vertical, as in Example 3.1.1. What deflections of joints A and B does this loading produce?*

The bar tensions have already been calculated (page 29), and are summarized in Figure 3.11(b). It helps to avoid mistakes if the calculation of the elongations is carried out in tabular form (Table 5.1). The construction of

TABLE 5.1

Calculation of bar elongations in Example 5.3.1

Bar	Length (mm)	Cross-section (mm^2)	E (kN/mm^2)	Tension (kN)	Elongation (mm)
AD	5000	500	210	50	2·38
AB	5000	500	210	86·6	4·12
BC	5000	500	210	86·6	4·12
BD	7070	1100	210	−122·5	−3·75

the displacement diagram follows exactly the procedure outlined above. It is shown in Figure 5.6. Notice that since e_{bd} is negative, point B in the structure moves closer to D, and so in the displacement diagram the component of *bd* parallel to BC is downwards and to the left (in contrast to Figure 5.5, which is plotted as if all elongations were positive). From the diagram

$$u_A = 2{\cdot}4 \text{ mm}, \qquad u_B = 4{\cdot}1 \text{ mm},$$

$$v_A = -13{\cdot}5 \text{ mm}, \qquad v_B = -9{\cdot}4 \text{ mm}.$$

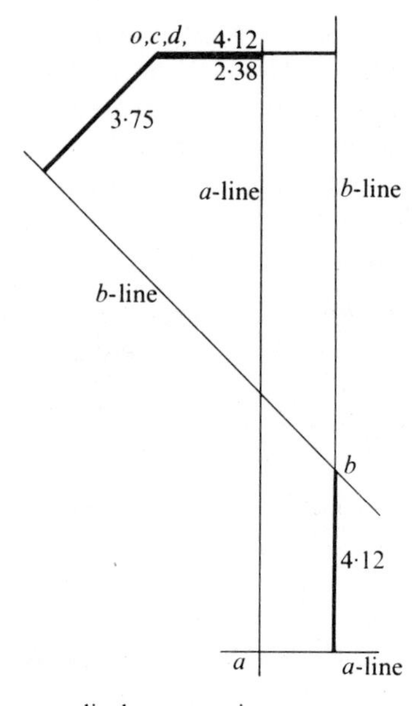

FIG. 5.6. Displacement diagram for framework in Example 5.3.1.

5.4. Bending and twisting deformations

Until now we have been concerned with deformations associated with the stretching of bars of frameworks. In more complicated kinds of structures, such as beams and frames, other kinds of deformation become important.

Imagine a straight beam, and draw a straight line to represent it (Figure 5.7(a)). Choose reference axes so that the 1-axis lies along the beam. Then suppose the beam to deflect into the new position shown in Figure 5.7(b). Each point in the beam moves in the 2-direction, and its displacement in that direction is v (consistent with the notation settled on in Section 5.1). Different points will have different displacements, and so the deflection v is a function of distance along the beam, located by a distance x measured from some arbitrary origin.

The deflected beam is no longer parallel to the 1-axis. Its inclination to it is measured by $\mathrm{d}v/\mathrm{d}x$, which is the tangent of the angle between the beam and the axis. In addition, the deflected beam is no longer straight: its curvature κ is the rate of change with distance along the beam of the angle between a tangent to the beam and a fixed direction. Analytic geometry shows it to be

$$\kappa = \frac{\dfrac{\mathrm{d}^2 v}{\mathrm{d}x^2}}{\left\{1+\left(\dfrac{\mathrm{d}v}{\mathrm{d}x}\right)^2\right\}^{3/2}}, \tag{5.4.1}$$

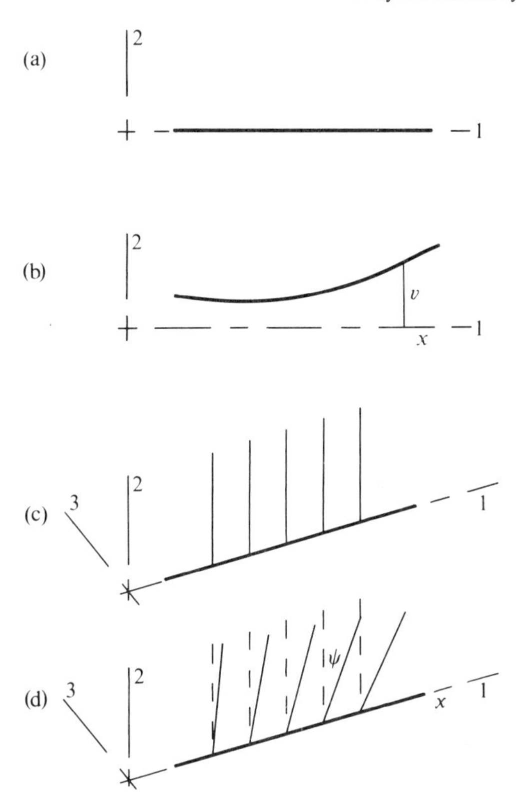

FIG. 5.7. (a) A straight beam, and reference axes. (b) Bending deformation of the beam. (c) and (d) Twisting deformation of the beam.

which is the general formula for the curvature of a curve defined in terms of rectangular coordinates. Its dimensions are $(\text{length})^{-1}$; it is the reciprocal of the radius of curvature.

It happens that this formula can almost always be simplified. In most real problems the inclination $\mathrm{d}v/\mathrm{d}x$ is sufficiently small for the second term in the denominator to be negligibly small compared to the first. Suppose, for instance, that $\mathrm{d}v/\mathrm{d}x$ is $0{\cdot}1$, which corresponds to a $5{\cdot}7°$ inclination, which is quite large. The denominator is then $1{\cdot}01^{3/2} = 1{\cdot}0150$, and differs from 1 by only $1{\cdot}5$ per cent. This suggests that we can normally replace eqn (5.4.1) by

$$\kappa = \frac{\mathrm{d}^2 v}{\mathrm{d}x^2}. \tag{5.4.2}$$

This makes the subsequent mathematics much easier. Except for very flexible structures, such as cables and suspended roofs, the approximation introduces very little error indeed, and from now on it will be assumed to hold.

Again, think of a straight beam which lies along the 1-axis. Imagine a series of pointers attached to the beam at different points, each one parallel to the 2-axis, so that they point upwards like the teeth of a comb (Figure 5.7(c)). Suppose that the beam now twists, as if under the action of a torque about the 1-axis. Each cross-section of the beam rotates about the 1-axis, and the pointers rotate with it (Figure 5.7(d)). If you find this hard to visualize, twist a comb, or take a pice of rubber tubing, stick pins into it at regular intervals so that all the pins are parallel, and then twist the tube. Each pointer rotates through an angle ψ. If ψ is zero, the beam is in its original position. If ψ is the same for all the pointers, the beam has just rotated about its axis, but has not twisted. Twisting causes a variation of ψ from one point to the next, and the rate at which ψ varies with x is a measure of the intensity of twist, measured by a twist

$$\omega = \frac{d\psi}{dx}. \tag{5.4.3}$$

5.5 Deflections of beams

When a beam is loaded in such a way that bending moments are set up, the beam bends. The curvature at a certain point on the beam is a function of the bending moment at the same point. The relationship between them depends on the dimensions of the beam cross-section, and on the material it is made of. How this relationship is derived is outside the scope of this book and belongs to the theory of solid mechanics and the strength of materials.

Often there is a linear relationship between the curvature κ and the bending moment M, so that

$$M = F\kappa.$$

The constant of proportionality F is called the flexural rigidity; its dimensions are (force)(length)^2.

In some beam problems the deflection can be found very simply. If the bending moment is known, it determines the curvature d^2v/dx^2, the second derivative of the displacement with respect to x. If this is integrated twice, it gives us the displacement. There are two integration constants to be found, but they are given by the support conditions. The following examples illustrate the technique.

Example 5.5.1. *A cantilever of length L carries a uniformly distributed downward load of intensity w per unit length (Figure 5.8). Derive an expression for the deflection of the cantilever under the load, if the bending moment is related to the curvature by*

$$M = F\kappa.$$

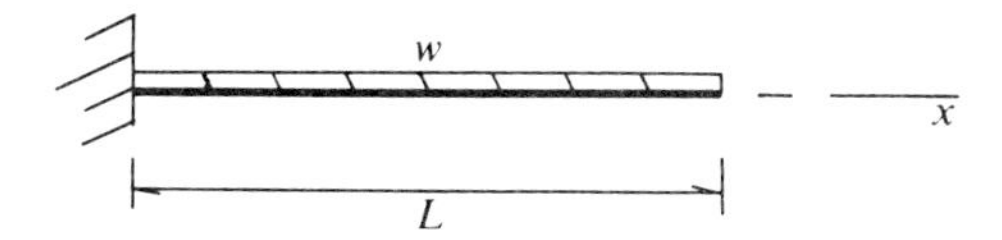

FIG. 5.8. Bending deformation of a cantilever.

The bending moment in the cantilever was calculated in Example 4.1.2, and is

$$M = -\tfrac{1}{2}w(L-x)^2 \tag{5.5.1}$$

at x from the left-hand end. Therefore

$$F\frac{\mathrm{d}^2v}{\mathrm{d}x^2} = -\frac{1}{2}w(L-x)^2. \tag{5.5.2}$$

Integrating with respect to x

$$F\frac{\mathrm{d}v}{\mathrm{d}x} = \frac{1}{6}w(L-x)^3 + c_1, \tag{5.5.3}$$

where c_1 is an integration constant. Integrating again,

$$Fv = -\frac{1}{24}w(L-x)^4 + c_1x + c_2, \tag{5.5.4}$$

and c_2 is a second integration constant. The integration conditions are obtained from the conditions at the supports. At the left-hand end of the cantilever, x is zero, and the deflection v is zero (because the cantilever is fixed to the foundation) and the inclination $\mathrm{d}v/\mathrm{d}x$ is zero (because the end is built-in, so that it cannot rotate). Setting x to zero in eqns (5.5.3) and (5.5.4) in turn, and using these conditions

$$0 = \frac{1}{6}wL^3 + c_1 \tag{5.5.5}$$

and

$$0 = \frac{1}{24}wL^4 + c_2, \tag{5.5.6}$$

and so

$$c_1 = -\frac{1}{6}wL^3 \tag{5.5.7}$$

and

$$c_2 = \frac{1}{24}wL^4. \tag{5.5.8}$$

Substituting back into (5.5.4),

$$v = -(w/F)\left\{\frac{1}{24}(L-x)^4 + \frac{1}{6}L^3x - \frac{1}{24}L^4\right\} \tag{5.5.9}$$

gives the deflection at any point on the cantilever.

Example 5.5.2. A uniform beam 14 m long rests on simple supports, and is loaded as shown in Figure 5.9. Its moment–curvature relationship is linear, with a flexural rigidity of 15 MN m^2. Find an expression for its deflection under the loads.

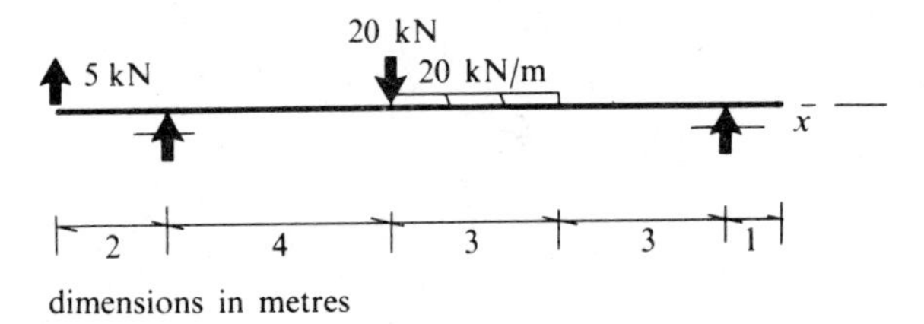

FIG. 5.9. Bending deformation of the beam in Example 5.5.2.

At any point on the beam the curvature is linearly proportional to the local bending moment. The way to find the bending moment was explained in Section 2.4, and the best notation in which to express it is the Macauley bracket notation of Example 4.1.3 (page 54). The support reactions are found by the method of Section 2.2, and are 33 kN at the left and 42 kN at the right. Since the flexural rigidity is 15 000 kN m^2,

$$15\,000\frac{d^2v}{dx^2} = \text{bending moment}$$

$$= 5[x+2] + 33[x] - 20[x-4] - 10[x-4]^2 + 10[x-7]^2 + 42[x-10], \tag{5.5.10}$$

where [ ] are Macauley brackets and x is measured from the left-hand support. There is one new detail which did not arise in Example 4.1.3. There the distributed load began 8 m from the left-hand end of the beam, and went on from there to the right-hand end. In this case, however, the distributed load begins 4 m from the left-hand support, at $x = 4$, and ends 7 m from it, at $x = 7$. We have to allow for this, but at the same time we naturally want to keep the simplicity of the Macauley notation. The way to do this is to think of the actual distributed load as the sum of two separate loadings:

(i) a downward distributed load of 20 kN/m, beginning at $x = 4$ and extending to the right-hand end of the beam, and

(ii) an upward distributed load of 20 kN/M, beginning at $x=7$ and extending to the right-hand end of the beam.

If we make an imaginary cut in the beam, as we learned to do in Section 4.1, and take moments about the cut, the contribution to the bending moment from loading (i) is $-10[x-4]^2$, and that from loading (ii) is $10[x-7]^2$, and so together they contribute

$$-10[x-4]^2+10[x-7]^2.$$

These terms appear in eqn (5.5.10). Although they look a little clumsy, this is in fact the most concise way of writing them, because if we tried to do without Macauley brackets we should need different expressions for the three ranges $x<4$ (to the left of the distributed load), $4<x<7$ (under the load), and $x>7$ (to the right of it).

Integrating eqn (5.5.10) once,†

$$15\,000\frac{dv}{dx}=\frac{5}{2}[x+2]^2+\frac{33}{2}[x]^2-10[x-4]^2-\frac{10}{3}[x-4]^3+\frac{10}{3}[x-7]^3 +21[x-10]^2+c_1. \quad (5.5.11)$$

Integrating a second time,

$$15\,000\, v=\frac{5}{6}[x+2]^3+\frac{11}{2}[x]^3-\frac{10}{3}[x-4]^3-\frac{5}{6}[x-4]^4+\frac{5}{6}[x-7]^4 +7[x-10]^3+c_1x+c_2, \quad (5.5.12)$$

and c_1 and c_2 are integration constants. At the left-hand support, $x=0$, the beam cannot deflect vertically, and so $v=0$; by substitution in (5.5.12),

$$0=\frac{20}{3}+c_2. \quad (5.5.13)$$

At the right-hand support $x=10$, and again $v=0$, and so

$$0=1440+5500-720-1080+67{\cdot}5+10c_1+c_2. \quad (5.5.14)$$

Solving these two equations

$$c_1=-520{\cdot}08$$
$$c_2=-6{\cdot}67 \quad (5.5.15)$$

Substitution back into (5.5.12) gives a complete expression for the displacement v at any point. Midway between the supports, for instance, x is 5 m,

† In the integration process, Macauley brackets are treated as single items, to be integrated whole, and not ones in which the individual terms within the bracket are integrated separately. The reason for this is that if instead we integrated terms within the bracket, we would lose the simple notation which tells us when a term is to be included and when it is to be set to zero.

and therefore

$$v=\frac{1}{15\,000}\left\{\frac{5}{6}(343)+\frac{11}{2}(125)-\frac{10}{3}(1)-\frac{5}{6}(1)-2600{\cdot}4-6{\cdot}67\right\}$$

$$=-0.109 \text{ m},$$

a downwards deflection as we should expect.

Example 5.5.3. A uniform beam is built-in to a rigid foundation at one end, and simply supported on a fixed support at the other end, as shown in Figure 5.10. Its flexural rigidity is F. It carries a load of intensity w per unit length, uniformly distributed along its whole length. Find an expression for its deflection.

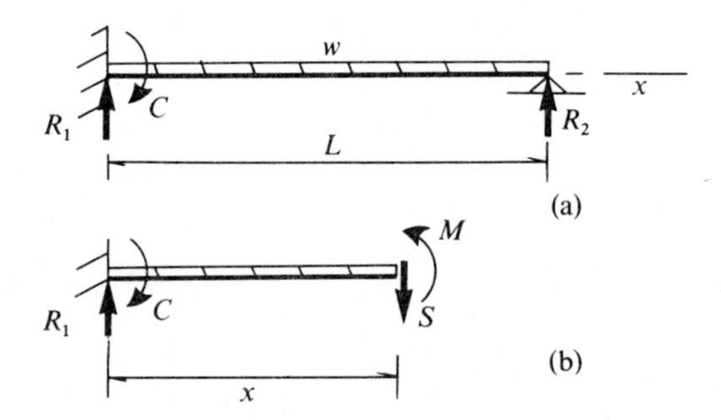

FIG. 5.10. Beam built-in at one end and simply supported at the other end.

If we follow the previous example, the first step is to find the support reactions. At the left-hand built-in end, there are a vertical reaction R_1 and a moment C, while at the right-hand simply supported end there is a vertical reaction R_2. If we try to find these reactions by the use of statics, we come up against the same difficulty as in Chapter 2, Example 2.3.3, which considered a framework resting on three supports. We can find only two independent equilibrium equations:

$$R_1+R_2-wL=0, \tag{5.5.16}$$

since the resulting vertical force on the beam is zero, and

$$-\tfrac{1}{2}wL^2+R_2L+C=0, \tag{5.5.17}$$

by taking moments about the right-hand support; but they are not enough to determine the three unknown quantities R_1, R_2, and C. The beam is statically indeterminate.

Statics alone can take us no further: we can only find the reactions by applying a geometric condition, that the reactions be such that although the beam deflects under the load, it does so in such a way that it still fits the supports. At the left-hand end, it is built-in, so that it neither deflects vertically nor rotates. At the right-hand end, it is simply supported, and does not deflect, though it can rotate. Some of the ways in which this beam might deflect are shown in Figure 5.11: mode (a) does not obey the conditions,

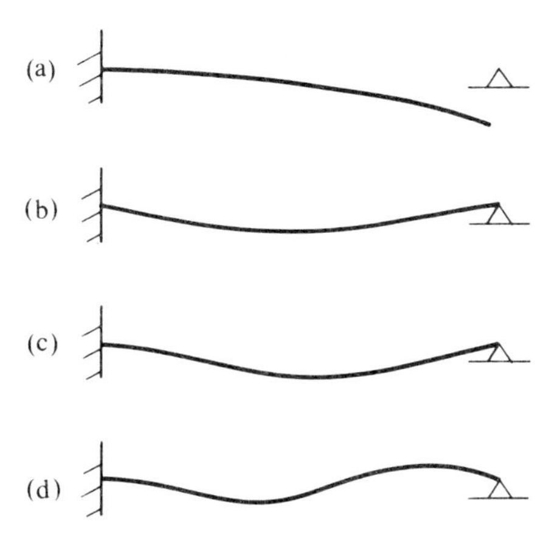

FIG. 5.11. Deflections of the beam in Figure 5.10.

because one end deflects below its support, and nor does mode (b), because the left-hand end rotates, but both modes (c) and (d) do obey the conditions. Geometric conditions like this are called compatibility conditions, and modes of deformation that obey them are said to be kinematically admissible.

Although we cannot yet find the values of R_1, R_2, and C, we can find the bending moment at a general point in terms of them. Making an imaginary cut x from the left-hand end (Figure 5.10(b)), and considering equilibrium of a segment to the left of the cut, in the usual way, the shear force S and bending moment M are

$$S = -wx + R_1 \tag{5.5.18}$$

$$M = -\tfrac{1}{2}wx^2 + R_1x + C. \tag{5.5.19}$$

In addition

$$F\frac{d^2v}{dx^2} = M,$$

where F is the flexural rigidity, and so

$$F\frac{d^2v}{dx^2} = -\frac{1}{2}wx^2 + R_1x + C. \tag{5.5.20}$$

Integrating twice,

$$F\frac{\mathrm{d}v}{\mathrm{d}x}=-\frac{1}{6}wx^3+\frac{1}{2}R_1x^2+Cx+c_1 \tag{5.5.21}$$

$$Fv=-\frac{1}{24}wx^4+\frac{1}{6}R_1x^3+\frac{1}{2}Cx^2+c_1x+c_2, \tag{5.5.22}$$

and c_1 and c_2 are integration constants. At the left-hand support, $x=0$, the deflection is zero, and there is no rotation, and so v and $\mathrm{d}v/\mathrm{d}x$ are both zero. Substituting into (5.5.21) and (5.5.22)

$$\begin{aligned}0&=c_1\\0&=c_2.\end{aligned} \tag{5.5.23}$$

We still have not used the geometric condition at the right-hand support, at $x=L$. There too there is no deflection, and so $v=0$; substituting into eqn (5.5.22)

$$0=-\frac{1}{24}wL^4+\frac{1}{6}R_1L^3+\frac{1}{2}CL^2. \tag{5.5.24}$$

This gives us a geometric condition relating the load to two of the unknowns. It has to be satisfied if the support conditions are obeyed. The statical conditions are of course obeyed as well, and they give us a second equation, eqn (5.5.17), which was got by taking moments. There are now two equations for two unknowns. Solving them as simultaneous equations,

$$\begin{aligned}R_1&=5wL/8\\C&=-wL^2/8,\end{aligned} \tag{5.5.25}$$

and, from eqn (5.5.16),

$$R_2=3wL/8, \tag{5.5.26}$$

and, substituting back into eqn (5.5.22),

$$v=\frac{1}{F}\left(-\frac{1}{24}wL^4+\frac{5}{48}wLx^3-\frac{1}{16}wL^2x^2\right). \tag{5.5.27}$$

This is the first statically indeterminate structure for which we have been able to find the forces set up by external loads. It was done by taking into account geometrical conditions as well as statical ones. In this instance the direct integration of the governing equation gave the solution, in a quite straightforward way, but generally it turns out to be better to use the virtual work method derived in the following chapter.

5.6. Problems

1. A triangular framework ABC is made from three bars, AB, BC, and CA, whose lengths are 1·5 m, 2·5 m, and 2 m respectively. Draw displacement diagrams for the following motions and deformations of the framework:

(i) a rotation about A, through $1{\cdot}5\times10^{-3}$ radians clockwise;
(ii) a rotation about the midpoint of AC, through $1{\cdot}5\times10^{-3}$ radians clockwise;
(iii) a deformation in which A does not move, AB remains in its initial direction, and AC increases in length by 2·5 mm;
(iv) a deformation in which A does not move, AB remains in its original direction, AC increases in length by 1 mm, and AB increases in length by 2 mm;
(v) a deformation in which A does not move, AB remains in its original direction, AC increases in length by 1 mm, AB increases in length by 2 mm, and BC decreases in length by 3 mm.

2. Figure 5.12 shows a pin-jointed framework composed of four bars, AB, BC, BD, and AD, whose nominal lengths are 2000 mm, 1000 mm, 2000 mm, and 2000 mm respectively. They are attached to a rigid foundation at C and D, which are exactly $1000\sqrt{3}$ mm apart.

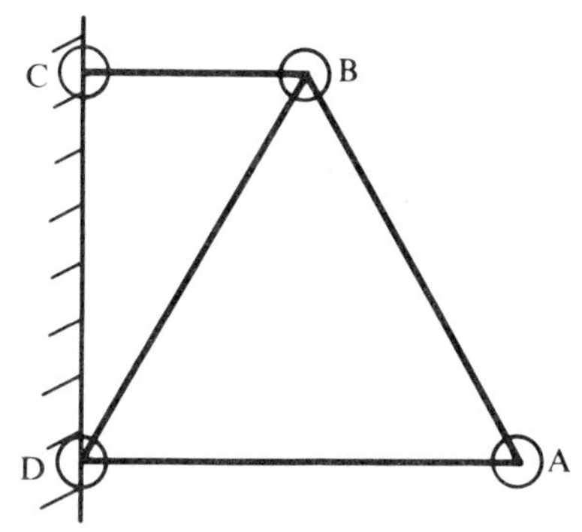

FIG. 5.12.

(i) BD is accidentally made 2 mm too long. When the framework is assembled, how far will A be from the location it would have if each bar had its nominal length?
(ii) Suppose instead that each bar has a manufacturing tolerance Δ, so that its true length is within Δ of its nominal length. The errors in different bars are not correlated, so that some may be Δ too long and others Δ too short. A design engineer wants to choose Δ so that he can be certain that when the framework is assembled A will not be more than 5 mm above or below its nominal position. What is the maximum value Δ can have?

3. A cantilever 3 m long carries an end load of 10 kN.

(i) If the flexural rigidity has the same value, 10 000 kN m^2, at all points on the cantilever, what is the end deflection under the load?
(ii) If the flexural rigidity F varies, because the cantilever is made deeper towards the built-in end, so that

$$F = 10\,000(x/3)^{\frac{1}{2}}\ \text{kN m}^2$$

at x m from the free end, what is then the end deflection?

(iii) If the relationship between bending moment and curvature is no longer linear, but instead

$$\kappa = \begin{cases} 0{\cdot}004 + 0{\cdot}0003M, & M < -20, \\ 0{\cdot}0001M, & -20 < M < 20, \\ -0{\cdot}004 + 0{\cdot}0003M, & M > 20, \end{cases}$$

where M is the bending moment in kN m and κ the curvature in m^{-1}, what is the end deflection?

Hint: if different relationships between M and κ apply in different parts of the cantilever, you will have to write different equations for d^2v/dx^2, integrate them separately, and then match the resulting equations by an appropriate choice of integration constants. What physical conditions will this matching choice of integration constants express?

4. A beam AB is 10 m long and simply supported at both ends. It carries a uniformly distributed load of intensity 100 kN/m, which extends over 4 m, from a point 5 m from A to a point 9 m from A. The uniform flexural rigidity of the beam is 100 MN m^2. Find

(i) the reactions at the supports,
(ii) expressions for the shear force and bending moment at a point at a distance x from A,
(iii) the location and magnitude of the maximum bending moment,
(iv) the location and magnitude of the maximum deflection.

5. A uniform girder 30 m long weighing 1200 kN rests on three jacks, A at one end, B 20 m from A, and C at the other end, 10 m from B. Initially the levels of the jacks are so adjusted that the weight is shared equally between jacks A and C; B is just in contact with the girder, but carries no load The flexural rigidity of the girder is 2500 MN m^2.

An engineer now wishes to lower jack C, so that it can be removed. How much must C be lowered before the reaction it exerts falls to zero? Jacks A and B do not move.

Hint: initially the three supports are not at the same level. The engineer is interested in the change in shape of the girder as the forces on it change. This change in shape is produced by the changes in the forces. Forces that act all the time, and do not change, do not produce any changes in shape.

6. A uniform beam is built-in at both ends. Its length is L and its uniform flexural rigidity F. Initially it is unloaded and transmits no bending moment. A uniformly distributed load wL is then applied to it. Decide what quantities you will use to describe the end reactions induced by the loading, find any statical relations between them (remembering that the beam and its loading are symmetric), and construct a general expression for the bending moment at a distance x from one end. Thence find the values of the reactions, and the central deflection of the beam.

7. A rail rests on a continuous foundation. It has a constant flexural rigidity F. The foundation has the property that if it is deflected through a distance v, it exerts a resisting force kv per unit length on the rail, in such a direction as to oppose the deflection. The rail carries a varying distributed load of w per unit length. Show that the deflection of the rail is governed by

$$F\frac{d^4v}{dx^4} + kv = w.$$

The behaviour of crane rails and railway track can be represented by this idealization, and it can be used to find the stresses set up in them. Although a railway line usually rests on distinct sleepers (cross-ties) at regular intervals, the sleepers are in practice close enough together for it to be reasonable to treat them as a continuous foundation when one is analysing track stresses. The equivalent foundation stiffness k is then an averaged value.

6. Virtual work

6.1. Introduction

THIS chapter introduces one of the most useful and powerful concepts in mechanics. It is called virtual work, and brings together concepts from both statics and geometry. It can be used to find the deflections of structures under loads, to find the forces set up within structures, and to derive general and useful statements about mechanics, such as the reciprocal theorem for linear structures and the limit theorems that we shall use to find collapse loads.

One kind of structure has particularly simple internal forces. This is the pin-jointed framework described in Chapter 3, in which all the bars carry axial tensions, and the only loads are concentrated forces applied at the joints. The statement of the virtual work theorem for pin-jointed frameworks is correspondingly simple, and will be derived first. A more general statement follows later. The theorem is quite general, and applies to any structure, determinate or indeterminate, elastic or inelastic, two-dimensional or three-dimensional.

6.2. Virtual work: pin-jointed frameworks

The following proof is for plane frameworks, but can easily be generalized to three dimensions.

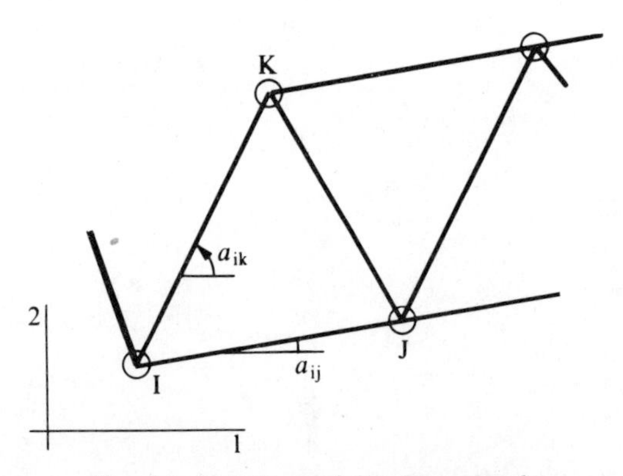

FIG. 6.1. Part of a pin-jointed framework.

Figure 6.1 shows a framework made up of straight bars hinged together. It lies in a vertical plane, and directions and force and displacement components will be described by reference to axes 1 and 2, the 1-axis being horizontal. The joints are identified by capital letters I, J, K . . . ; Bow's notation is not being used here. The orientation of a bar is described by the angle it makes with the horizontal axis. Angle a_{ij} is the angle between the positive 1-axis and a line from I to J, measured counter-clockwise positive (Figure 6.1). Consistently, a_{ji} is the angle between the 1-axis and a line from J to I, and therefore

$$a_{ji} = a_{ij} + 180°. \tag{6.2.1}$$

If the structure moves or distorts, its joints move, and the bars remain connected together at the joints. The notation for displacement is the u, v notation introduced in Chapter 5, and a subscript denotes which joint is being referred to, so that u_i is the horizontal component of the displacement of joint I, v_i the vertical component of the same displacement, u_j the horizontal component of the displacement of joint J, and so on. Concentrate attention on a single bar IJ. It remains straight as the framework distorts. Figure 6.2 is a displacement diagram for the movements of the ends of the

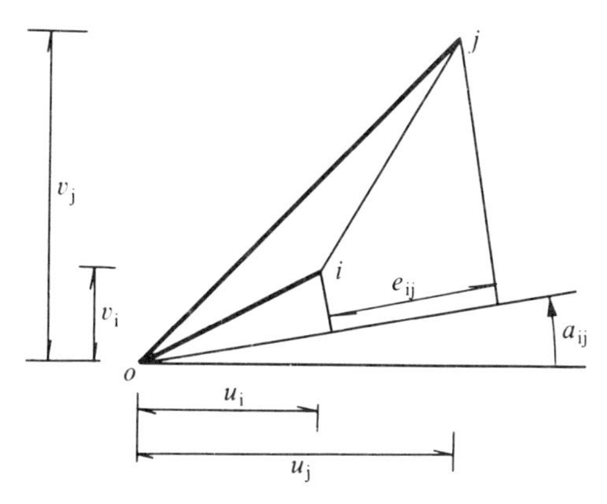

FIG. 6.2. Displacement diagram for joints I and J in Figure 6.1.

bar. The displacement of I is *oi*, whose components are u_i and v_i; the displacement of J is *oj*, whose components are u_j and v_j. The relative displacement between I and J is *ij*, whose components are $u_j - u_i$ and $v_j - v_i$. We found in Chapter 5 that the elongation e_{ij} of a bar connecting I to J is the component parallel to the bar itself of the relative displacement between the ends, the perpendicular component being accounted for by rotation. In the displacement diagram, e_{ij} is the component of *ij* at a_{ij} to the horizontal, and straightforward trigonometry shows it to be

$$e_{ij} = (u_j - u_i) \cos a_{ij} + (v_j - v_i) \sin a_{ij}. \tag{6.2.2}$$

Since a_{ij} and a_{ji} differ by 180°,

$$\begin{aligned} \sin a_{ij} &= -\sin a_{ji} \\ \cos a_{ij} &= -\cos a_{ji}, \end{aligned} \tag{6.2.3}$$

and the expression for e_{ij} can be rewritten

$$e_{ij} = -(u_i \cos a_{ij} + u_j \cos a_{ji} + v_i \sin a_{ij} + v_j \sin a_{ji}). \tag{6.2.4}$$

You will notice that we have continued with the assumption discussed in Chapter 5, that rotations are small because joint displacements are much smaller than the lengths of the bars. The purely geometric result expressed by eqn (6.2.4) applies to any small deformation, no matter how it is produced. It will be put on one side while we reconsider the statics of the framework. Before doing so, we need an elementary mathematical result.

Multipliers

Consider an equation

$$a = b + c. \tag{6.2.5}$$

If we multiply the whole equation by a multiplier λ, we get

$$\lambda a = \lambda b + \lambda c, \tag{6.2.6}$$

a new equation, which holds for any value of λ, from which the original equation can be recovered by putting λ equal to 1. In the same way, a second equation

$$d = e + f \tag{6.2.7}$$

can be multiplied by another multiplier μ, giving

$$\mu d = \mu e + \mu f. \tag{6.2.8}$$

The multiplied equations can be added together

$$\lambda a + \mu d = \lambda b + \lambda c + \mu e + \mu f. \tag{6.2.9}$$

This equation holds for any value of λ and any value of μ. The multipliers are quite independent of one another. If we want, we can get back the original equations by giving the multipliers in eqn (6.2.9) particular values. If, for example, we put λ equal to zero and μ equal to 1, we get back eqn (6.2.7). We can also find combinations of the original equations: if we put

$$\lambda = 1$$

$$\mu = -3$$

into eqn (6.2.9), we have

$$a - 3d = b + c - 3e - 3f, \tag{6.2.10}$$

which could instead have been derived by multiplying eqn (6.2.7) by three and then subtracting the result from eqn (6.2.5).

The possibility of exploiting multipliers in this way will be used in the proof of the theorem.

Statics

Imagine the framework to be loaded in some way. Figure 6.3 shows joint I. It is subjected to two kinds of forces, those exerted by external loads

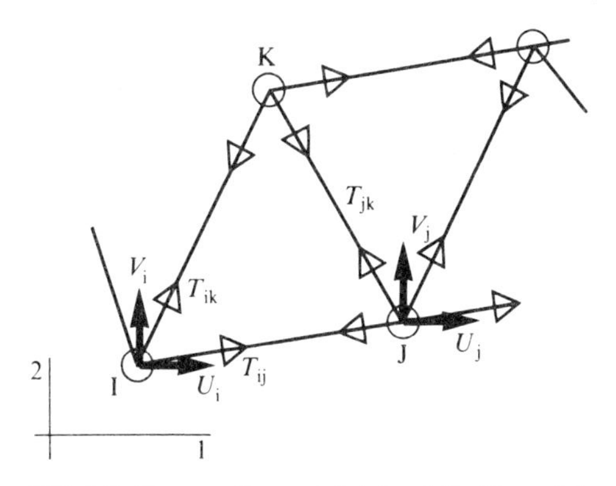

FIG. 6.3. Forces acting on joints I and J in Figure 6.1.

applied at the joint itself, and those exerted by the tensions in the bars that meet there. The external load at I has a horizontal component U_i and a vertical component V_i. Each bar has a tension. The tension in bar IJ is T_{ij}, and the horizontal component of the force it exerts on I is $T_{ij} \cos a_{ij}$, and the vertical component is $T_{ij} \sin a_{ij}$. Similarly, the tension in bar IK is T_{ik}, and the horizontal and vertical components of the force it exerts on I are $T_{ik} \cos a_{ik}$ and $T_{ik} \sin a_{ik}$, whereas the components of the force it exerts on K, at the other end of the bar, are $T_{ki} \cos a_{ki}$ and $T_{ki} \sin a_{ki}$. Since the bar has only one tension, T_{ik} and T_{ki} are equal, but, as we saw in eqn (6.2.1), a_{ki} and a_{ik} are not equal and differ by 180°.

Since joint I is in equilibrium, the horizontal components of the resulting force on it must be zero, and so

$$U_i + \sum_{\substack{\text{all bars}\\ \text{meeting at I}}} T_{ij} \cos a_{ij} = 0, \tag{6.2.11}$$

where the summation is over all the bars that are connected to joint I. Similarly

$$V_i + \sum_{\substack{\text{all bars}\\ \text{meeting at I}}} T_{ij} \sin a_{ij} = 0. \tag{6.2.12}$$

These are nothing more than joint equilibrium equations, like the equations written in Example 3.1.2, but now written so as to apply to any framework.

Now let the first of these equilibrium equations be multiplied by an arbitrary multiplier m_i, and the second be multiplied by another multiplier n_i; then

$$U_i m_i + \sum_{\substack{\text{all bars}\\ \text{meeting at I}}} T_{ij} m_i \cos a_{ij} = 0 \tag{6.2.13}$$

$$V_i n_i + \sum_{\substack{\text{all bars}\\ \text{meeting at I}}} T_{ij} n_i \sin a_{ij} = 0. \tag{6.2.14}$$

The two equations can be added together:

$$U_i m_i + V_i n_i + \sum_{\substack{\text{all bars}\\ \text{meeting at I}}} T_{ij}(m_i \cos a_{ij} + n_i \sin a_{ij}) = 0, \tag{6.2.15}$$

and this equation holds for any m_i and any n_i, just as eqn (6.2.9) held for any λ and any μ. Similar equations can be written for each of the joints of the framework: for joint K, for instance,

$$U_k m_k + V_k n_k + \sum_{\substack{\text{all bars}\\ \text{meeting at K}}} T_{kj}(m_k \cos a_{kj} + n_k \sin a_{kj}) = 0. \tag{6.2.16}$$

Notice that different multipliers are used for each joint.

We now carry out a second summation. Equation (6.2.15) describes the equilibrium of joint I, eqn (6.2.16) describes the equilibrium of joint K, and there is one of these equations for each of the joints. If we add all the equations together, in this way carrying out a summation over all the joints, we have

$$\sum_{\text{all joints I}} (U_i m_i + V_i n_i) + \sum_{\text{all joints I}} \left(\sum_{\substack{\text{all bars}\\ \text{meeting at I}}} T_{ij}(m_i \cos a_{ij} + n_i \sin a_{ij}) \right) = 0. \tag{6.2.17}$$

The double summation that forms the second term in this equation contains contributions from all the bars. If we pick out the contribution from one particular bar IJ, it is

$$T_{ij} m_i \cos a_{ij} + T_{ij} n_i \sin a_{ij} + T_{ji} m_j \cos a_{ji} + T_{ji} n_j \sin a_{ji},$$

the first two terms coming from the equilibrium equation for joint I, and the last two from the equilibrium equation for joint J. Since T_{ij} and T_{ji} are equal, this can be rewritten

$$T_{ij}(m_i \cos a_{ij} + m_j \cos a_{ji} + n_i \sin a_{ij} + n_j \sin a_{ji}).$$

The contributions of other bars to the overall sum are similar; that for bar IK, for instance, is

$$T_{ik}(m_i \cos a_{ik} + m_k \cos a_{ki} + n_i \sin a_{ik} + n_k \sin a_{ki}).$$

It follows that we can rewrite eqn (6.2.17) as

$$\sum_{\text{all joints I}} (U_i m_i + V_i n_i) + \sum_{\text{all bars IJ}} T_{ij}(m_i \cos a_{ij} + m_j \cos a_{ji} + n_i \sin a_{ij} + n_j \sin a_{ji}) = 0. \tag{6.2.18}$$

So far we have not used any concepts but statics, and some algebra, the fact that we can multiply through any equation by an arbitrary multiplier. Until now the multipliers m_i and n_i for each equilibrium equation have been numbers, without any physical significance. At this point in the argument they are given particular values: the multiplier in the horizontal equilibrium equation for joint I is set equal to the horizontal displacement of the same joint in the deformation considered at the beginning of this section, so that

$$m_i = u_i, \tag{6.2.19}$$

the multiplier of the vertical equilibrium equation for the same joint is set equal to the corresponding vertical displacement

$$n_i = v_i, \tag{6.2.20}$$

and so on for all the joints

$$\begin{aligned} m_j &= u_j, \\ n_j &= v_j, \\ m_k &= u_k, \end{aligned} \tag{6.2.21}$$

and so on. If we make these substitutions into eqn (6.2.18), we get

$$\sum_{\text{all joints I}} (U_i u_i + V_i v_i) + \sum_{\text{all bars IJ}} T_{ij}(u_i \cos a_{ij} + u_j \cos a_{ji} + v_i \sin a_{ij} + v_j \sin a_{ji}) = 0. \tag{6.2.22}$$

Now compare the expression that multiplies T_{ij} in this equation with the geometric result expressed in eqn (6.2.4), which was derived earlier when we were considering deformations. It is simply $-e_{ij}$, the negative extension of bar IJ, and so eqn (6.2.22) is

$$\sum_{\text{all joints I}} (U_i u_i + V_i v_i) = \sum_{\text{all bars IJ}} T_{ij} e_{ij}. \tag{6.2.23}$$

This is a remarkable result. It relates two independent groups of quantities, one of them concerned with geometry, a set of joint displacements u_i and v_i which are geometrically compatible with certain bar elongations e_{ij}, the other concerned with statics, a set of joint loads U_i and V_i in equilibrium with certain bar tensions T_{ij}. The only thing the sets have in common is that they relate to the same framework. It is important to understand that it has not been assumed that the elongations and displacements are those produced by the bar forces. The two sets are independent: one satisfies comparibility, the other equilibrium, and that is all.

The horizontal displacement u_i corresponds to the horizontal external load U_i, and the work that would be done by U_i if the joint I moved horizontally through u_i is $U_i u_i$, consistent with the conventional definition of 'work' in mechanics. The phrase 'virtual work' underlines the fact that the joint need not actually move in this way. Similarly, the displacement v_i corresponds to the load V_i, and the work of V_i on v_i would be $V_i v_i$. One can think of the left-hand side of eqn (6.2.23) as the 'external' virtual work of the loads on their corresponding displacements, and in the same way think of the right-hand side as the 'internal' virtual work of the bar tensions acting through their corresponding bar elongations. Any more direct physical interpretation is likely to be confusing, and to lead to the risk of losing sight of the fact that the loads and deformations are independent.

We now consider a simple example of the use of virtual work. Figure 6.4(a) shows a statically determinate framework made up of two bars AB and AB pinned to a rigid foundation. It carries a vertical load W. The tensions set up in the two bars could of course be found in the usual way, by considering the equilibrium of joint A, but instead they can be found by virtual work.

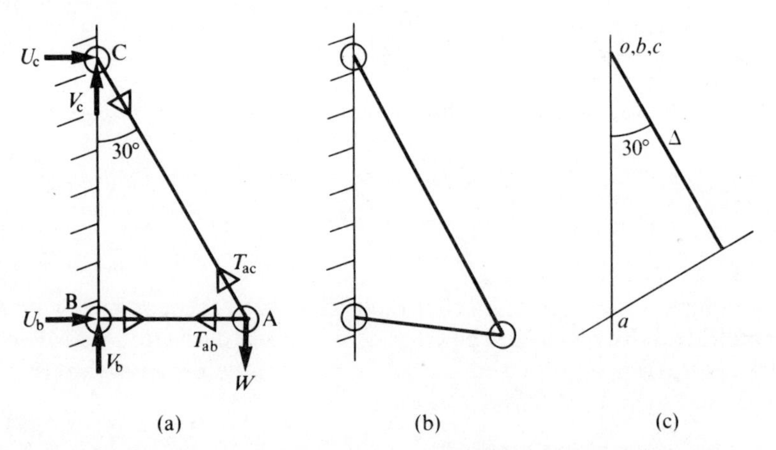

FIG. 6.4. (a) Forces in a pin-jointed framework. (b) Deformation corresponding to an elongation of bar AC. (c) Displacement diagram.

Suppose we want to find the tension T_{ac}. In the right-hand side of eqn (6.2.23) the bar tensions in the equilibrium set of forces are multiplied by the corresponding elongations in the compatible set of displacements. We want the actual tension induced by the load W, and so must choose the equilibrium set of loads and tensions to be the actual set corresponding to W. We still have a free choice of compatible displacements and elongations. If we choose a set of displacements which have a non-zero elongation for the actual bar we are interested in, but zero elongation for the other bar, then the only term on the right-hand side of eqn (6.2.23) will be the tension T_{ac} multiplied by the corresponding extension, and we shall have an equation for T_{ac}. Therefore, choose the displacements so that

bar AC increases in length by Δ,
bar AB remains the same length, and
joints B and C do not move.

The corresponding deformation is shown in an exaggerated form in Figure 6.4(b), and the displacement diagram in 6.4(c). Joints B and C do not move, so they are represented by a single point coincident with o. Bar AB remains the same length, and so A can only move vertically downward, and the point representing it lies vertically below b. The elongation e_{ac} is Δ, and so Δ is the component of ac on the displacement diagram in the direction of AC, at 30° to the vertical. On the diagram, therefore, a must lie $2\Delta/\sqrt{3}$ below b, and that is the vertical displacement of A. The left-hand side of eqn (6.2.23) has six terms; two for each joint. They can be written out in a table, for the actual forces and for the displacements we have just chosen, recalling the sign convention that forces and displacements to the right and upward are positive.

TABLE 6.1

Joint	Force U_i or V_i	Displacement u_i or v_i	Product U_iu_i or V_iv_i
A horizontal	0	0	0
A vertical	$-W$	$2\Delta/\sqrt{3}$	$2W\Delta/\sqrt{3}$
B horizontal	U_b	0	0
B vertical	V_b	0	0
C horizontal	U_c	0	0
C vertical	V_c	0	0

$$\sum_{\text{all joints}} (U_iu_i + V_iv_i) = 2W\Delta/\sqrt{3}$$

The right-hand side has two terms, one for each bar, and they are calculated in Table 6.2.

TABLE 6.2

Bar	Tension T_{ij}	Elongation e_{ij}	Product $T_{ij}e_{ij}$
AB	T_{ab}	0	0
AC	T_{ac}	Δ	$T_{ac}\Delta$
		$\sum_{\text{all bars}}$	$T_{ij}e_{ij} = T_{ac}\Delta$

Therefore the theorem tells us that

$$2W\Delta/\surd 3 = T_{ac}\Delta, \tag{6.2.24}$$

and so

$$T_{ac} = 2W/\surd 3, \tag{6.2.25}$$

which is the same result as we get directly from joint equilibrium.

In this example virtual work has been used to solve a very simple problem in statics, and has done this by converting it into a geometrical problem for a carefully chosen set of displacements and elongations obeying a compatibility condition. The geometry is embodied in the displacement diagram. In the next section, virtual work is used the other way round, to solve a geometric problem by using statics.

6.3. Deflections of pin-jointed frameworks

Imagine that we know, or know how to calculate, the elongations of the bars of a framework, and that we want to find the deflections of the joints that result. In Chapter 5 we solved problems like this by constructing displacement diagrams. It is often easier to use a method based on virtual work, which has the added advantage that it can be used for three-dimensional frameworks.

The virtual work statement is about two independent sets of quantities. On the one hand, there is a set of joint displacements, and corresponding bar elongations, which obey a geometric compatibility condition. On the other hand, there is a set of forces in equilibrium with certain bar tensions. Each set can be chosen independently of the other. In this instance, we know the actual bar elongations, and want the actual joint deflections, and so it must be them that we choose as the compatible set. Having done that, we can still choose the set of forces in equilibrium as we wish, and we do so so that the deflections we are interested in appear in the 'external' virtual work

$$\sum_{\text{all joints I}} (U_i u_i + V_i v_i),$$

but the other deflections do not appear, because they are multiplied by zero.

In Section 5.3 we set out to find the displacements of the joints of the framework illustrated in Figure 6.5(a). Joints C and D do not move, the bars

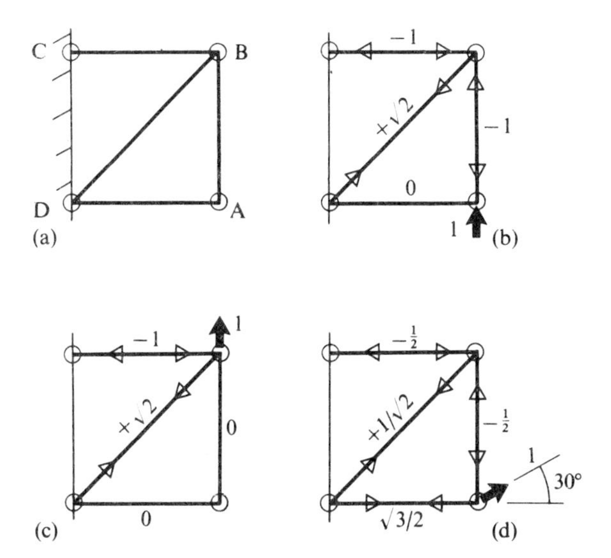

FIG. 6.5. (a) Pin-jointed framework ABCD. (b) Bar tensions corresponding to unit vertical external load at A. (c) Bar tensions corresponding to unit vertical external load at B. (d) Bar tensions corresponding to a unit external load at A, directed at 30° to the horizontal.

have elongations e_{ab}, e_{ad}, e_{bc}, and e_{bd}, and we wanted to find the joint displacements u_a, v_a, u_b, and v_b. The same problem will now be solved by virtual work. We choose the actual elongations and displacements as the compatible set.

Suppose we want to find v_a. Imagine a unit vertical load to act externally on the framework at A; the load is in the direction of the displacement we want to find, and acts at the same point. Such a load would induce tensions in the bars, and support reactions at C and D. Denote the tensions induced by this load by T^*_{ij}, and the reactions by U^*_c, V^*_c, U^*_d, and V^*_d. Their values are shown in Figure 6.5(b). Choose these loads and tensions, which do not actually act on the framework, but would be in equilibrium if they did, as the equilibrium set in the virtual work statement, eqn (6.2.23).

In summary, then, we apply virtual work to

(i) the actual elongations e_{ij} and displacements u_i and v_i, but to

(ii) imaginary bar tensions T^*_{ij} and loads U^*_i and V^*_i, corresponding to unit vertical load at A.

Applying eqn (6.2.23)

$$\sum_{\text{all joints I}} (U^*_i u_i + V^*_i v_i) = \sum_{\text{all bars}} T^*_{ij} e_{ij}. \qquad (6.3.1)$$

The summations are carried out in Tables 6.3 and 6.4.

TABLE 6.3

Joint	U^*_i	u_i	$U^*_i u_i$	V^*_i	v_i	$V^*_i v_i$	$U^*_i u_i + V^*_i v_i$
A	0	u_a	0	1	v_a	v_a	v_a
B	0	u_b	0	0	v_b	0	0
C	U^*_c	0	0	V^*_c	0	0	0
D	U^*_d	0	0	V^*_d	0	0	0
					$\sum_{\text{all joints}}$	$(U^*_i u_i + V^*_i v_i) = v_a$	

TABLE 6.4

Bar	Tension T^*_{ij}	Elongation e_{ij}	$T^*_{ij} e_{ij}$
AB	-1	e_{ab}	$-e_{ab}$
AD	0	e_{ad}	0
BC	-1	e_{bc}	$-e_{bc}$
BD	$+\sqrt{2}$	e_{bd}	$e_{bd}\sqrt{2}$
		$\sum_{\text{all bars}}$	$T^*_{ij} e_{ij} = e_{bd}\sqrt{2} - e_{bc} - e_{ab}$

The external virtual work is simply v_a, the product of the unit load and the displacement we are looking for. Substituting into eqn (6.3.1)

$$v_a = e_{bd}\sqrt{2} - e_{bc} - e_{ab}, \qquad (6.3.2)$$

which is the result obtained by a quite different technique in Section 5.3. Like that result, it is purely geometric: the use in the virtual work statement of the imaginary set of forces in Figure 6.5(b) was simply an artifice through which the geometric relationship was derived. Note that the elongations are not those that the imaginary force at A would produce if it actually loaded the framework, and indeed the result applies however the elongations are produced, whether by loading or by some other effect such as thermal expansion.

In order to find a different displacement, a different imaginary load is used, which naturally induces a different set of bar tensions. The imaginary load has unit magnitude and the direction of the required displacement. Thus, to find the vertical deflection at B, suppose a unit vertical load at B (Figure 6.5(c)), and call the corresponding bar tensions T_{ij}^{**}. Again apply virtual work, and choose the actual elongations and displacements as the compatible set, and now choose as the equilibrium set the unit load at B, the corresponding reactions at C and D, and the bar tensions T_{ij}^{**}. Then

$$v_b = \sum_{\text{all bars}} T_{ij}^{**} e_{ij}, \tag{6.3.3}$$

which is computed in Table 6.5, and gives

$$v_b = e_{bd}\sqrt{2} - e_{bc}. \tag{6.3.4}$$

TABLE 6.5

Bar	Tension T_{ij}^{**}	Elongation e_{ij}	$T_{ij}^{**} e_{ij}$
AB	0	e_{ab}	0
AD	0	e_{ad}	0
BC	-1	e_{bc}	$-e_{bc}$
BD	$+\sqrt{2}$	e_{bd}	$e_{bd}\sqrt{2}$
			$\sum_{\text{all bars}} T_{ij}^{**} e_{ij} = e_{bd}\sqrt{2} - e_{bc}$

If instead we wanted to find the component at 30° to the horizontal of the deflection of A, the imaginary unit load to use would be a unit load in the same direction, which in turn would induce a third set of bar tensions T_{ij}^{***} (Figure 6.5(d)). Using this third load system, the deflection at 30° to the horizontal is

$$\sum_{\text{all bars}} T_{ij}^{***} e_{ij}.$$

Example 6.3.1. The bridge truss shown in Figure 6.6 rests on a simple support at A and a roller support at E. The elongation of its bars under load follows Hooke's law. It is made from steel, elastic modulus 210 kN/mm². The vertical and diagonal bars have cross-sections of 3000 mm², and the horizontals have a cross-section of 5000 mm².

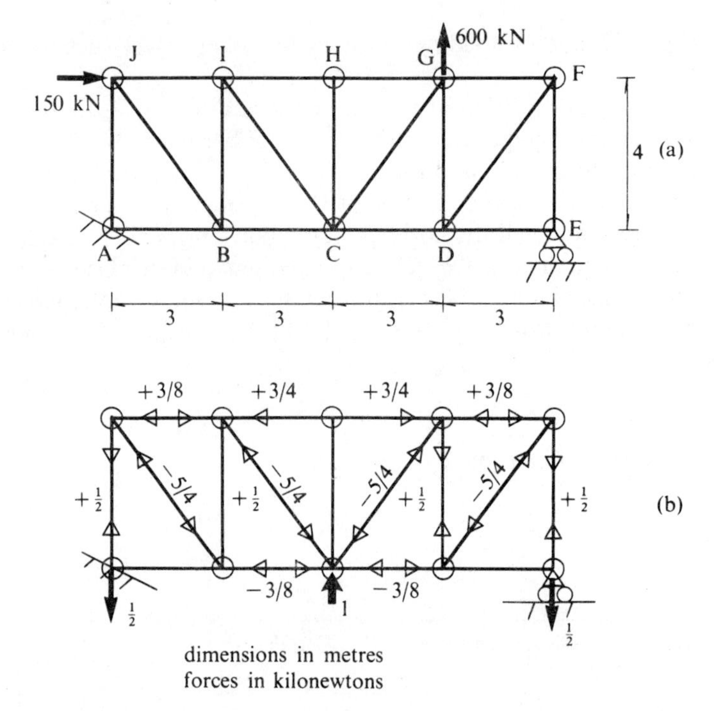

FIG. 6.6. (a) Bridge truss under load. (b) Bar tensions corresponding to unit vertical external load at C.

(i) *Find the vertical deflection of C under the loading shown.*

(ii) *If a bridge is built so that its lower chord (ABCDE in this case) is horizontal, it appears to sag. This optical illusion, that straight horizontal lines appear to bend downwards, was known to the ancient Greeks, and in the design of important buildings their architects compensated for it by making lines which they wanted to appear horizontal actually curve slightly upwards. In the same way, bridges are often cambered upwards. An engineer wishes to camber the bridge truss in Figure 6.6 so as to raise joint C 50 mm above a straight line between A and E, and to do this by increasing the length of each of the bars of the upper chord (FGHIJ) slightly above its nominal length of 3000 mm. What should the increases in length be?*

(i) The tensions in the bars under this loading were calculated in Example 3.1.2 (page 31), and are listed in Table 6.7. The elongations are calculated from

$$e_{ij} = \frac{(\text{tension } T_{ij})(\text{length of bar IJ})}{(\text{cross-section})(\text{elastic modulus})}, \tag{6.3.5}$$

which expresses Hooke's law. To find the vertical deflection of C, let an imaginary unit vertical load act at C. The corresponding bar tensions are T^*_{ij}, listed in Figure 6.6(b) and found in the customary way. Once again, choose the actual elongations e_{ij} and corresponding displacements as the compatible set in the application of virtual work, and choose the unit load and tensions T^*_{ij} of Figure 6.6(b) as the equilibrium set. The external virtual work is calculated in Table 6.6

TABLE 6.6

Joint	U^*_i	u_i	V^*_i	v_i	$(U^*_i u_i + V^*_i v_i)$
A	0	0	−1/2	0	0
B	0	u_b	0	v_b	0
C	0	u_c	+1	v_c	v_c
D	0	u_d	0	v_d	0
E	0	u_e	−1/2	0	0
F	0	u_f	0	v_f	0
G	0	u_g	0	v_g	0
H	0,	u_h	0	v_h	0
I	0	u_i	0	v_i	0
J	0	u_j	0	v_j	0
					v_c

This gives

$$v_c = \sum_{\text{all bars}} T^*_{ij} e_{ij}. \tag{6.3.6}$$

The summation is carried out in Table 6.7. Writing the calculation out in tabular form like this helps to avoid mistakes. The elongations e_{ij} in the fifth column are calculated by eqn (6.3.5), using the elastic modulus of 210 kN/mm^2 specified in the question.

Accordingly

$$v_c = 8{\cdot}8 \text{ mm}.$$

(ii) Imagine a 'perfect' reference framework which has exactly the dimensions shown in Figure 6.6(a), so that the verticals are 4000 mm long, the horizontals 3000 mm, and the diagonals 5000 mm. Now let certain bars elongate: the joints then move, and the framework takes up a new configuration. The deformation from the perfect reference configuration to the final actual configuration obeys the compatibility conditions. Call the associated joint displacements and elongations u^{**}_i, v^{**}_i, and e^{**}_{ij}, and select them as the geometrically compatible set. Since we are interested in designing bar elongations so that v^{**}_c takes a certain value, choose as the equilibrium set the set of tensions and external loads corresponding to an imaginary unit

TABLE 6.7

Bar	Tension T_{ij} (kN)	Length (mm)	Area (mm^2)	e_{ij} (mm)	T^*_{ij}	$T^*_{ij}e_{ij}$ (mm)
AB	+150	3000	5000	0·428	0	0
BC	0	3000	5000	0	−3/8	0
CD	−300	3000	5000	−0·857	−3/8	0·321
DE	0	3000	5000	0	0	0
JI	0	3000	5000	0	3/8	0
IH	+150	3000	5000	0·428	3/4	0·321
HG	+150	3000	5000	0·428	3/4	0·321
GF	+300	3000	5000	0·857	3/8	0·321
AJ	+200	4000	3000	1·270	1/2	0·635
BI	+200	4000	3000	1·270	1/2	0·635
CH	0	4000	3000	0	0	0
DG	+400	4000	3000	2·540	1/2	1·270
EF	+400	4000	3000	2·540	1/2	1·270
BJ	−250	5000	3000	−1·984	−5/8	1·240
CI	−250	5000	3000	−1·984	−5/8	1·240
CG	+250	5000	3000	1·984	−5/8	−1·240
DF	−500	5000	3000	−3·968	−5/8	2·480
						8·814

load at C, as in Figure 6.6(b). Applying the virtual work equation, eqn (6.2.23), with these choices

$$\sum_{\text{all joints}} (U^*_i u^{**}_i + V^*_i v^{**}_i) = \sum_{\text{all bars}} T^*_{ij} e^{**}_{ij}. \tag{6.3.7}$$

The external virtual work is simply v^{**}_c, from Table 6.6, and so

$$v^{**}_c = \sum_{\text{all bars}} T^*_{ij} e^{**}_{ij}. \tag{6.3.8}$$

TABLE 6.8

Bar	T^*_{ij}	e^{**}_{ij}	$T^*_{ij}e^{**}_{ij}$
FG	+3/8	Δ	(3/8)Δ
GI	+3/4	Δ	(3/4)Δ
HI	+3/4	Δ	(3/4)Δ
IJ	+3/8	Δ	(3/8)Δ
remaining bars	not needed	0	0
			(9/4)Δ

The required 50 mm displacement of C is to be obtained by making equal changes in length in the four bars FG, GH, HI, and IJ of the upper chord. Let this required change in length be Δ. Table 6.8 presents the calculation of the right-hand side of eqn (6.3.8); substituting into that equation,

$$50 = (9/4)\Delta,$$

and so

$$\Delta = 22{\cdot}2 \text{ mm}.$$

Three-dimensional frameworks

The statement of virtual work for plane frameworks needs only a slight change to make it applicable to three-dimensional frameworks. Denote the 3-component of the displacement of joint I by w_i, consistent with the notation of Chapter 5, and the corresponding component of the external load by W_i. Together they form an external virtual work term $W_i w_i$, and the generalization of eqn (6.2.23) becomes

$$\sum_{\text{all joints}} (U_i u_i + V_i v_i + W_i w_i) = \sum_{\text{all bars}} T_{ij} e_{ij}. \tag{6.3.9}$$

Example 6.3.2. Nine identical bars are hinged together to make the space frame illustrated in Figure 6.7. It has the form of two regular tetrahedra, ABCD and ABCE, with one common face. Each bar obeys Hooke's law, and has the same length L and cross-section A. Equal and opposite forces Q are applied to D and E. By how much does the distance between D and E increase?

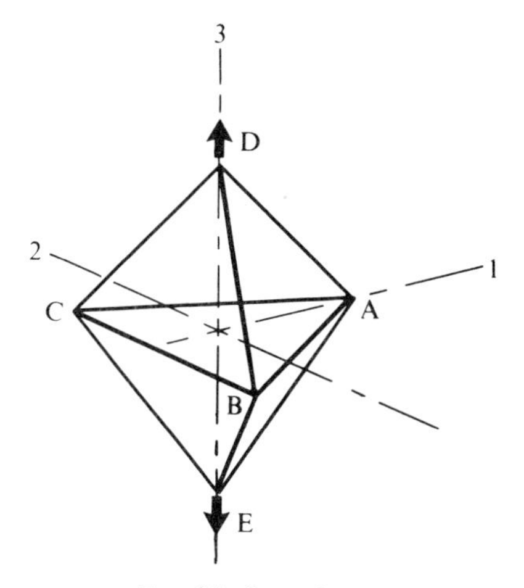

FIG. 6.7. Space frame.

Choose the reference axes shown, so that the origin is the midpoint of DE and the 1-axis passes through A. The framework has three-fold symmetry about DE, and is symmetrical with respect to the plane ABC. Coordinates of the joints are located by the following position vectors

$$\begin{aligned}
&\text{A} \quad \mathbf{r}_a = (L/\sqrt{3})\mathbf{i} + 0\mathbf{j} + 0\mathbf{k} \\
&\text{B} \quad \mathbf{r}_b = -(L/2\sqrt{3})\mathbf{i} - (L/2)\mathbf{j} + 0\mathbf{k} \\
&\text{C} \quad \mathbf{r}_c = -(L/2\sqrt{3})\mathbf{i} + (L/2)\mathbf{j} + 0\mathbf{k} \\
&\text{D} \quad \mathbf{r}_d = 0\mathbf{i} + 0\mathbf{j} + (\sqrt{2}/\sqrt{3})\mathbf{k} \\
&\text{E} \quad \mathbf{r}_e = 0\mathbf{i} + 0\mathbf{j} - (\sqrt{2}/\sqrt{3})\mathbf{k}.
\end{aligned} \tag{6.3.10}$$

Unit vectors in the directions of BA and AD are respectively

$$\begin{aligned}
\frac{\mathbf{r}_a - \mathbf{r}_b}{|\mathbf{r}_a - \mathbf{r}_b|} &= \frac{\sqrt{3}}{2}\mathbf{i} + \tfrac{1}{2}\mathbf{j} + 0\mathbf{k}, \\
\frac{\mathbf{r}_d - \mathbf{r}_a}{|\mathbf{r}_d - \mathbf{r}_a|} &= -\frac{1}{\sqrt{3}}\mathbf{i} + 0\mathbf{j} + \frac{\sqrt{2}}{\sqrt{3}}\mathbf{k}.
\end{aligned} \tag{6.3.11}$$

Because of the symmetry of the framework and its loading

$$\begin{aligned}
&T_{ab} = T_{bc} = T_{ac}, \\
&T_{ad} = T_{ae} = T_{bd} = T_{be} = T_{cd} = T_{ce}.
\end{aligned} \tag{6.3.12}$$

Using this symmetry, and considering the equilibrium of joint D, resolving forces in the 3-direction,

$$T_{ad} = Q/\sqrt{6}. \tag{6.3.13}$$

Considering next the equilibrium of joint A, resolving forces in the 1-direction,

$$0 = 2\frac{1}{\sqrt{3}}T_{ad} + 2\frac{\sqrt{3}}{2}T_{ab},$$

and so

$$T_{ab} = -(2/3\sqrt{6})Q. \tag{6.3.14}$$

The elongations e_{ij} can then be calculated from the bar tensions.

Joints D and E both move vertically. In order to find their relative displacement, imagine a unit upward vertical load applied at D, and a unit downward vertical load applied at E. Call the corresponding tensions T^*_{ij}. Apply virtual work in the usual way, choosing T^*_{ij} and the unit loads as the equilibrium force set and the actual displacements and elongations as the geometrically compatible set. Then

$$v_d - v_e = \sum_{\text{all bars}} T^*_{ij} e_{ij}. \quad (6.3.15)$$

In fact T^*_{ij} here correspond to the actual bar tensions T_{ij}, with Q set equal to unity. This is not so in general, as we saw in the last example, but is so in this instance because the actual loading corresponds to the relative displacement we want. The calculation is carried out in Table 6.9 and the required relative displacement is $(11/9)QL/AE$.

TABLE 6.9

Bar	T_{ij}	e_{ij}	T^*_{ij}	$T^*_{ij}e_{ij}$	Number of similar bars	$T^*_{ij}e_{ij}$ all similar bars
AB	$-\frac{2}{3\sqrt{6}}Q$	$-\frac{2}{3\sqrt{6}}\frac{QL}{AE}$	$\frac{-2}{3\sqrt{6}}$	$\frac{2}{27}\frac{QL}{AE}$	3	$\frac{2}{9}\frac{QL}{AE}$
AD	$\frac{Q}{\sqrt{6}}$	$\frac{1}{\sqrt{6}}\frac{QL}{AE}$	$\frac{1}{\sqrt{6}}$	$\frac{1}{6}\frac{QL}{AE}$	6	$\frac{QL}{AE}$
						$\frac{11}{9}\frac{QL}{AE}$

6.4. Forces in statically indeterminate frameworks

We saw in Section 3.2 that there are some frameworks for which the equilibrium conditions by themselves do not determine the bar tensions. There are too many bars for the framework to be statically determinate, and we need to make use of additional geometric conditions. The conditions we use express the fact that the bars are still connected together after the framework has deformed under load, so that the elongations of the bars must 'fit' the displacements of the joints. In other words, one or more compatibility conditions have to be satisfied. These conditions can be expressed either explicitly, through the calculation and matching of displacements, or implicitly, through the use of virtual work.

Alternative approaches to this problem are best discussed in terms of a specific example.

Example 6.4.1. A square pin-jointed framework ABCD (Figure 6.8) is attached to a rigid foundation at C and D. All the bars obey Hooke's law, and have the same elastic modulus E. Bars AB, BC, and DA have cross-sectional area A, and bars AC and BD have cross-sectional area $\frac{1}{2}$A. Initially, before the framework is loaded, the tensions in the bars are all zero. A vertical load P is then applied at A. What tensions are induced in the bars?

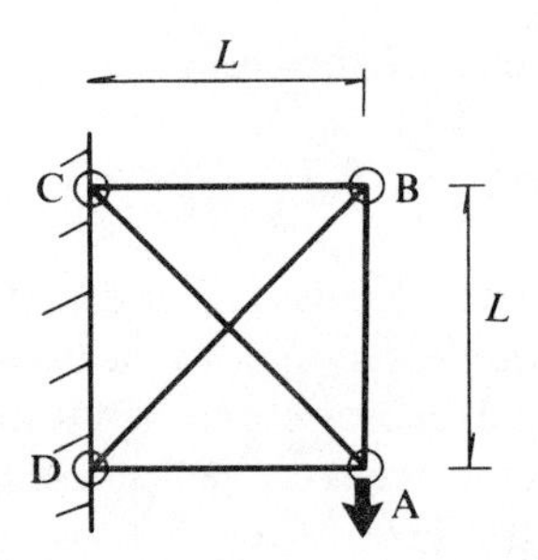

FIG. 6.8. A statically indeterminate pin-jointed framework.

In Example 3.2.1 an attempt was made to find the forces within this framework by statics. The attempt failed, because the framework is statically indeterminate. It has just one bar too many for it to be determinate. If one of the bar tensions could be found, the others could be found by statics: we say that the framework is once redundant.

The problem will be solved by two distinct methods. In the first method the compatibility condition is used directly. One thinks of an imaginary cut in the structure, and finds the tensions that must be set up in the bars if the two sides of the cut are just to meet. The relative displacements required are found by drawing displacement diagrams. The second method depends on the virtual work principle. The forces set up in the framework are such that the elongations of different bars are compatible with zero relative displacements at the joints. In the virtual work equation the displacements and elongations are chosen as the actual displacements and elongations, and a careful choice of an imaginary equilibrium system of loads and tensions then gives an equation from which the actual bar tensions can be found.

Solution by displacement-diagram method

Imagine that before the structure is loaded a narrow cut is made in bar AB, just above joint A. Label the upper side of the cut E, so that BE is the severed bar that formerly joined B to A. The cut makes the framework statically determinate. When the load is applied at A, the tensions in the bars of the modified structure are those shown in Figure 6.9(a). As a result, bar AC elongates, and bar AD shortens, while the remaining bars do not alter in length (because their tensions are zero). The corresponding elongations are

$$\begin{aligned} e_{ac} &= +4PL/AE \\ e_{ad} &= -PL/AE \qquad (6.4.1) \\ e_{bc} &= e_{bd} = e_{be} = 0. \end{aligned}$$

The corresponding displacement diagram is plotted in Figure 6.9(b). Joint A

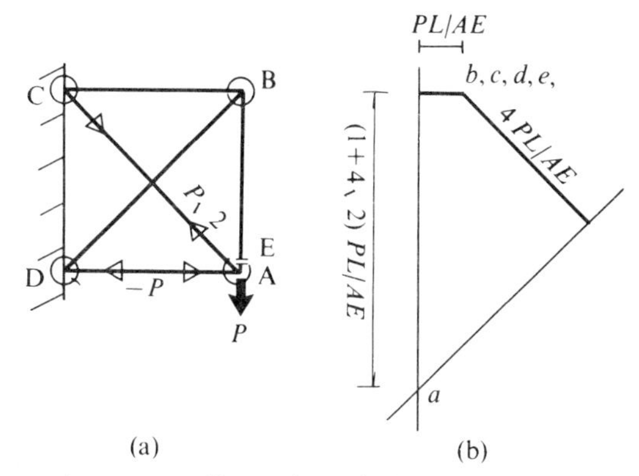

FIG. 6.9. (a) Bar tensions present if a cut is made in AB just above A. (b) Corresponding displacement diagram.

has moved downwards with respect to E, so that the cut has widened, by $(1+4\sqrt{2})PL/AE$, from the geometry of the displacement diagram.

In reality, of course, there is no cut in AB. Instead a tension develops in it: one can think of it as the tension that would have to be applied between the sides of the cut to keep them from moving apart, so that A and E had the same displacement. Call this induced tension R; as yet its magnitude is unknown. The tensions in the other bars of the framework are then those shown in Figure 6.10(a). They can all be calculated in terms of P and R by the usual method, the equilibrium of joint A giving the tensions in AC and AD, and that of joint B giving the tensions in BC and BD. The corresponding elongations are worked out in Table 6.10.

TABLE 6.10

Bar	Length	Area	Tension T_{ij}	Elongation e_{ij}
BC	L	A	R	RL/AE
BE	L	A	R	RL/AE
AD	L	A	$-(P-R)$	$-(P-R)L/AE$
AC	$L\sqrt{2}$	$\frac{1}{2}A$	$(P-R)\sqrt{2}$	$4(P-R)L/AE$
BD	$L\sqrt{2}$	$\frac{1}{2}A$	$-R\sqrt{2}$	$-4RL/AE$

We can now find the condition that has to be obeyed if A and E are to have the same displacement. Construct displacement diagrams to find the displacements of A and E individually, in terms of P and R, and then impose the condition that they must match, so that the faces of the imaginary cut neither separate nor overlap. Figure 6.10(b) is a displacement diagram, constructed in the usual way, that locates A. From its geometry, the

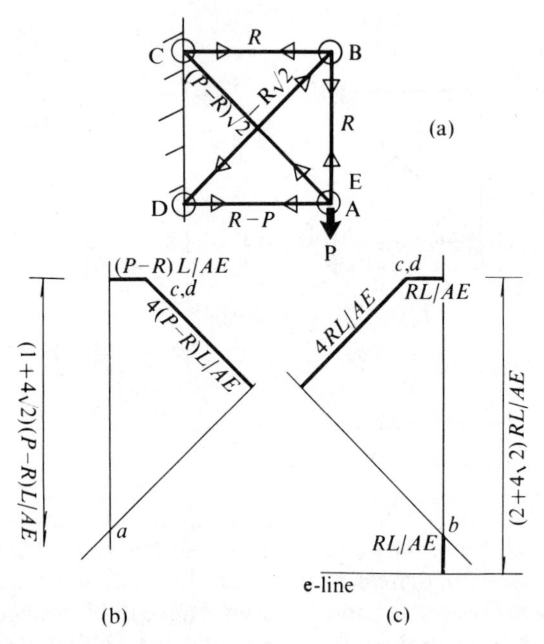

FIG. 6.10. (a) Bar tensions present if AB is not cut. (b) Displacement diagram for joint A. (c) Displacement diagram for point E.

downward displacement of A is $(1+4\sqrt{2})(P-R)L/AE$. The displacement diagram for E is shown separately in Figure 6.10(c). Displacement b is located by the elongations of BC and BD, and since the elongation of BE is RL/AE, the point e representing the displacement of E lies on a line RL/AE below b. From the geometry of the diagram, the downward displacement of E is $(2+4\sqrt{2})RL/AE$.

Since A and E must in the actual structure have the same displacement,

$$(1+4\sqrt{2})(P-R)L/AE=(2+4\sqrt{2})RL/AE, \qquad (6.4.2)$$

and therefore

$$R=\frac{1+4\sqrt{2}}{3+8\sqrt{2}}P, \qquad (6.4.3)$$

and the compatibility condition gives us the unknown bar tension R. The other bar tensions can now be found by substituting back into Table 6.10 (column 4), or Figure 6.10(a).

This method has been included here because of the emphasis it puts on the compatibility condition. It is not usually the best way of solving such a problem. As in the calculation of deflections, it is generally better to use the virtual work method.

Solution by virtual work method

The framework is once redundant. Choose one of the bars, AB, and call the unknown tension in it R. Next find the tensions in all the other bars in terms of R and the external load. It is usually best to do this by superposing two simpler tension systems:

(i) the tensions that would be set up if the external load were present but R were zero, and

(ii) the tensions that would be set up if R were present but there were no external load.

These systems of tensions are shown separately in Figure 6.11(a) and (b), and together in Figure 6.11(c), which shows the actual bar tensions (and of

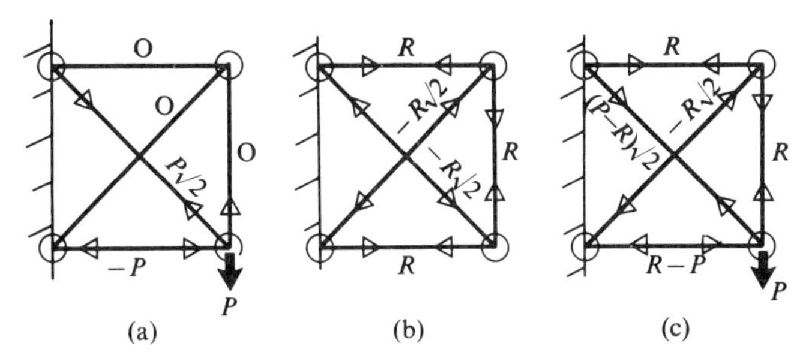

FIG. 6.11. (a) Bar tensions set up if external load acts but redundant tension is zero. (b) Bar tensions set up if external load is absent but redundant tension R is present. (c) Actual bar tensions when external load and redundant tension R are both present.

course agrees with Figure 6.10(a)). The elongation e_{ij} of each bar can now be calculated, in terms of R and the external load P. The virtual work equation tells us that in general

$$\sum_{\text{all joints}} (U_i^* u_i^{**} + V_i^* v_i^{**}) = \sum_{\text{all bars}} T_{ij}^* e_{ij}^{**}, \qquad (6.2.23)$$

where T_{ij}^*, U_i^*, and V_i^* are an equilibrium set of tensions and loads, and e_{ij}^{**}, u_i^{**}, and v_i^{**} an independent set of compatible elongations and displacements. We identify this latter compatible set of elongations and displacements with the actual elongations and displacements under the actual tensions and loads. However, we choose the equilibrium set to be an *imaginary* one that is in equilibrium with zero external loads, except at support points where the actual displacements are constrained to be zero. In symbols, we choose tensions T_{ij}^* and loads U_i^* and V_i^* so that

$$U_i^* = V_i^* = 0 \qquad \text{at all joints i}, \qquad (6.4.4)$$

unless the conditions of the problem constrain u_i to be zero (in which case

U_i^* need not be zero at that joint) or v_i to be zero (in which case V_i^* need not be zero at that joint).

For such a system of bar tensions the left-hand side of eqn (6.2.23) vanishes identically, so that

$$0 = \sum_{\text{all bars}} T_{ij}^* e_{ij}^{**}. \tag{6.4.5}$$

In the present problem a set of bar tensions satisfying eqn (6.4.4) has to be in equilibrium with zero external loads at A and B. Already we have such a set. It is shown in Figure 6.11(b), and corresponds to the bar tensions that would exist if R were present but there were no external loads (except at C and D, which are fixed). If, for simplicity, we replace R by 1, we get the system of bar tensions shown in Figure 6.12, which meets the conditions we want. Such

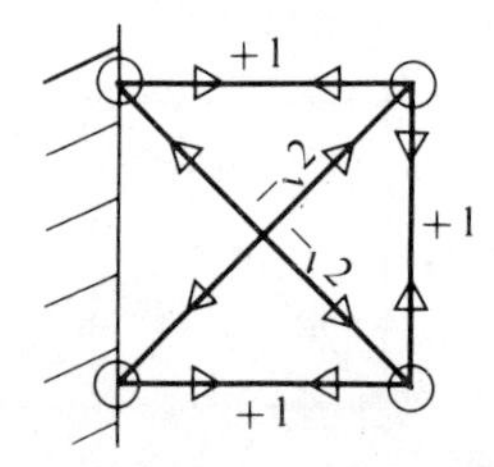

FIG. 6.12. Self-equilibrating system of bar tensions.

a system of tensions and reactions is called self-equilibrating. We now choose this system as T_{ij}^* in eqn (6.4.5), evaluate the summation, and equate the sum to zero. The summation is carried out in Table 6.11.
Accordingly, from eqn (6.4.5)

$$0 = (3+8\sqrt{2})RL/AE - (1+4\sqrt{2})PL/AE$$
$$R = \frac{1+4\sqrt{2}}{3+8\sqrt{2}}P, \tag{6.4.6}$$

which agrees with the previous result.

TABLE 6.11

Bar	Length	Area	Tension T_{ij}	Elongation e_{ij}	T_{ij}^*	$T_{ij}^* e_{ij}$
BC	L	A	R	RL/AE	1	RL/AE
AB	L	A	R	RL/AE	1	RL/AE
AD	L	A	$-(P-R)$	$-(P-R)L/AE$	1	$-(P-R)L/AE$
AC	$L\sqrt{2}$	$\frac{1}{2}A$	$(P-R)\sqrt{2}$	$4(P-R)L/AE$	$-\sqrt{2}$	$-4\sqrt{2}(P-R)L/AE$
BD	$L\sqrt{2}$	$\frac{1}{2}A$	$-R\sqrt{2}$	$-4RL/AE$	$-\sqrt{2}$	$4RL\sqrt{2}/AE$
						$\{(3+8\sqrt{2})R-(1+4\sqrt{2})P\}\frac{L}{AE}$

It is important to understand that the geometric condition that has been applied here is the same as the one applied explicitly in the displacement diagram method. We are not using any different physical principle. The choice of an imaginary set of bar tensions was a device to enable us to make use of virtual work, which combines equilibrium and geometrical concepts. Exactly the same device was used in Section 6.3 to find displacements.

In statically determinate frameworks the bar tensions are independent of the way in which the elongation of an individual bar depends on its tension. In statically indeterminate frameworks this is no longer true: we made use of the Hooke's law relationship between tension and elongation in calculating column 5 of Table 6.11 and column 5 of Table 6.10. If the relationship had been different, the value of R would have been different. Suppose, for instance, that the relationship between the tension in AC and its elongation had instead been

$$\frac{\text{elongation}}{\text{length}} = c_1 T_{ac} + c_2 (T_{ac})^2, \tag{6.4.7}$$

where c_1 and c_2 are constants. The elongation e_{ac} would then have been

$$\{c_1(P-R)\sqrt{2} + 2c_2(P-R)^2\}L$$

and the entry in the final column of Table 6.11 corresponding to AC would have been

$$-\{c_1(P-R)\sqrt{2} + 2c_2(P-R)^2\}L\sqrt{2}.$$

The condition expressed by eqn (6.4.5) would then have been

$$0 = (3+4\sqrt{2})RL/AE - PL/AE - \{c_1(P-R)\sqrt{2} + 2c_2(P-R)^2\}L\sqrt{2} \tag{6.4.8}$$

instead of eqn (6.4.6), and R, the solution of this equation, would have had a different value.

Example 6.4.2. The framework shown in Figure 6.13 carries a load of 240 kN at C. All its bars are made of the same material, which obeys Hooke's law; the

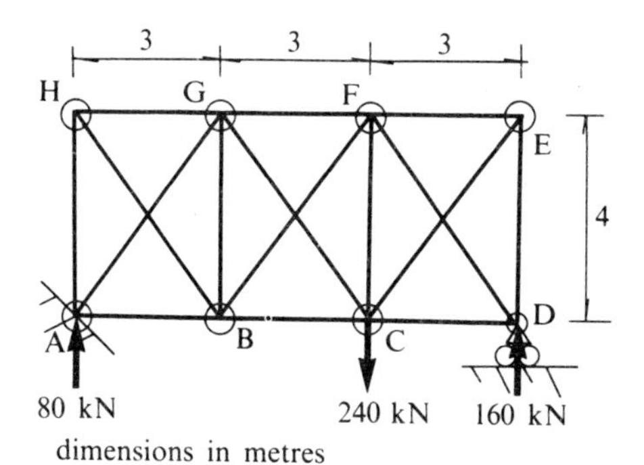

FIG. 6.13. Statically indeterminate framework.

cross-sectional areas are 2000 mm² for the verticals, 5000 mm² for the horizontals, and 2500 mm² for the diagonals. What tensions does the load induce in the bars?

The external reactions on the framework can be found by statics, and are shown in Figure 6.13.

The framework is three times redundant. If one bar is removed in each panel, it becomes determinate; if any more are removed, it becomes a mechanism. Choose the tensions in the diagonal bars AG, BF, and CE as the unknown redundant tensions, in terms of which all the bar tensions can be determined, and call them Q, R, and S. The first step is to find the tensions in the bars of the framework in terms of the external loads and the redundants. It is easiest to find those tensions as the sums of those that correspond to the four following simpler systems:

(i) external load present, Q, R, and S all zero,
(ii) Q present, no external load, R and S both zero,
(iii) R present, no external load, Q and S both zero, and
(iv) S present, no external load, Q and R both zero.

These systems are shown individually in Figure 6.14. The bar tensions T_{ij} are listed in Table 6.12 (page 108), and the elongations e_{ij} are calculated in column 5 of the same table.

The virtual work method will be used. Since there are three unknown redundant tensions, we are going to need three compatibility equations like eqn (6.4.6). In Example 6.4.1 the framework was once redundant, and there was one set of self-equilibrating bar tensions in equilibrium with zero external loads (except perhaps at points which were known to have no displacements in the actual framework). That suggests that in the present problem we should look for sets of bar tensions that obey the same conditions, and that we should expect to find three of them.

As in Example 6.4.1, suitable self-equilibrating sets of tensions are those obtained by setting the external loads equal to zero, one of the redundant tensions equal to one, and the remaining redundant tensions equal to zero. One of them corresponds to system (ii) above, with $Q=1$, $R=S=0$, and no external loads. Its bar tensions are denoted $T_{ij}^{(1)}$. If in the virtual work equation (6.2.23) we choose the equilibrium set of tensions and external loads as this self-equilibrating set, and choose as the elongations and joint displacements the actual elongations e_{ij} and displacements u_i and v_i, we have

$$0 = \sum_{\text{all bars}} T_{ij}^{(1)} e_{ij}. \tag{6.4.9}$$

A second self-equilibrating set of tensions $T_{ij}^{(2)}$ corresponds to (iii) above, with $R=1$, $Q=S=0$, and again no external loads. Applying virtual work again, still with the actual elongations and displacements, but now with this

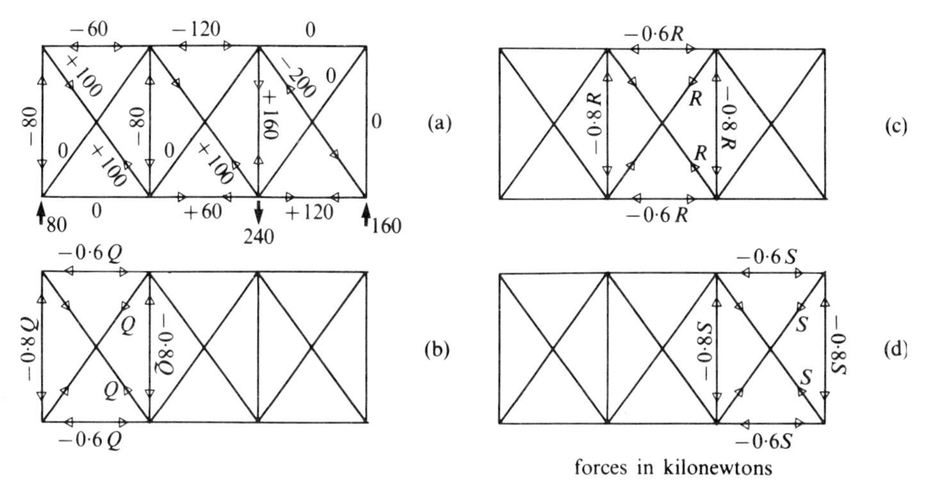

FIG. 6.14. (a) Bar tensions set up if external loads are present, and redundant tensions Q, R, and S are all zero. (b) Bar tensions set up if external loads are absent, and R and S are zero, but Q is not. (c) Bar tensions set up if external loads are absent, and Q and S are zero, but R is not. (d) Bar tensions set up if external loads are absent, and Q and R are zero, but S is not.

different equilibrium set of tensions $T_{ij}^{(2)}$, we have

$$0 = \sum_{\text{all bars}} T_{ij}^{(2)} e_{ij}. \tag{6.4.10}$$

Finally, a third self-equilibrating set $T_{ij}^{(3)}$ corresponds to (iv) above, with $S = 1$, $Q = R = 0$, and no external loads. Applying virtual work a third time

$$0 = \sum_{\text{all bars}} T_{ij}^{(3)} e_{ij}. \tag{6.4.11}$$

The three equations are independent. The summations are carried out in Table 6.12, and give

$$0 = 477{\cdot}6 + 6{\cdot}992Q + 1{\cdot}28R \tag{6.4.12}$$

$$0 = 93{\cdot}6 + 1{\cdot}28Q + 6{\cdot}992R + 1{\cdot}28S \tag{6.4.13}$$

$$0 = -699{\cdot}2 + 1{\cdot}28R + 6{\cdot}992S, \tag{6.4.14}$$

which can readily be solved, and give

$$\begin{aligned} Q &= -64 \text{ kN} \\ R &= -21 \text{ kN} \\ S &= 104 \text{ kN}. \end{aligned} \tag{6.4.15}$$

TABLE 6.12

Bar	Length (mm)	Area (mm^2)	Tension (kN)	Elongation (mm) $\times 1/E$	$T_{ij}^{(1)}$	$T_{ij}^{(1)} e_{ij}$ (mm) $\times 1/E$	$T_{ij}^{(2)}$	$T_{ij}^{(2)} e_{ij}$ (mm) $\times 1/E$	$T_{ij}^{(3)}$	$T_{ij}^{(3)} e_{ij}$ (mm) $\times 1/E$
AB	3000	5000	$-0{\cdot}6Q$	$-0{\cdot}36Q$	$-0{\cdot}6$	$0{\cdot}216Q$	0	0	0	0
GH	3000	5000	$-60-0{\cdot}6Q$	$-36-0{\cdot}36Q$	$-0{\cdot}6$	$21{\cdot}6+0{\cdot}216Q$	0	0	0	0
BC	3000	5000	$60-0{\cdot}6R$	$36-0{\cdot}36R$	0	0	$-0{\cdot}6$	$-21{\cdot}6+0{\cdot}216R$	0	0
FG	3000	5000	$-120-0{\cdot}6R$	$-72-0{\cdot}36R$	0	0	$-0{\cdot}6$	$43{\cdot}2+0{\cdot}216R$	0	0
CD	3000	5000	$120-0{\cdot}6S$	$72-0{\cdot}36S$	0	0	0	0	$-0{\cdot}6$	$-43{\cdot}2+0{\cdot}216S$
EF	3000	5000	$-0{\cdot}6S$	$-0{\cdot}36S$	0	0	0	0	$-0{\cdot}6$	$0{\cdot}216S$
AH	4000	2000	$-80-0{\cdot}8Q$	$-160-1{\cdot}6Q$	$-0{\cdot}8$	$128+1{\cdot}28Q$	0	0	0	0
BG	4000	2000	$-80-0{\cdot}8Q-0{\cdot}8R$	$-160-1{\cdot}6Q-1{\cdot}6R$	$-0{\cdot}8$	$128+1{\cdot}28Q$ $+1{\cdot}28R$	$-0{\cdot}8$	$128+1{\cdot}28Q$ $+1{\cdot}28R$	0	0
CF	4000	2000	$160-0{\cdot}8R-0{\cdot}8S$	$320-1{\cdot}6R-1{\cdot}6S$	0	0	$-0{\cdot}8$	$-256+1{\cdot}28R$ $+1{\cdot}28S$	$-0{\cdot}8$	$-256+1{\cdot}28R$ $+1{\cdot}28S$
DE	4000	2000	$-0{\cdot}8S$	$-1{\cdot}6S$	0	0	0	0	$-0{\cdot}8$	$1{\cdot}28S$
AG	5000	2500	Q	$2Q$	1	$2Q$	0	0	0	0
BH	5000	2500	$100+Q$	$200+2Q$	1	$200+2Q$	0	0	0	
BF	5000	2500	R	$2R$	0	0	1	$2R$	0	0
CG	5000	2500	$100+R$	$200+2R$	0	0	1	$200+2R$	0	0
CE	5000	2500	S	$2S$	0	0	0	0	1	$2S$
DF	5000	2500	$-200+S$	$-400+2S$	0	0	0	0	1	$-400+2S$
						$477{\cdot}6$ $+6{\cdot}992Q+1{\cdot}28R$		$93{\cdot}6$ $+1{\cdot}28Q+6{\cdot}992R$ $+1{\cdot}28S$		$-699{\cdot}2$ $+1{\cdot}28R+6{\cdot}992S$

6.5. Problems

1. Repeat Problem 2 in Section 5.6, but now use the virtual work method.

2. The pin-jointed framework in Figure 6.15 is made from 11 identical bars, for each of which the elongation is linearly related to the tension by

$$\text{elongation} = K(\text{tension}).$$

The framework is simply supported at A and D, and carries vertical loads W at B and C. Find the resulting vertical deflection at B and horizontal deflection at D.

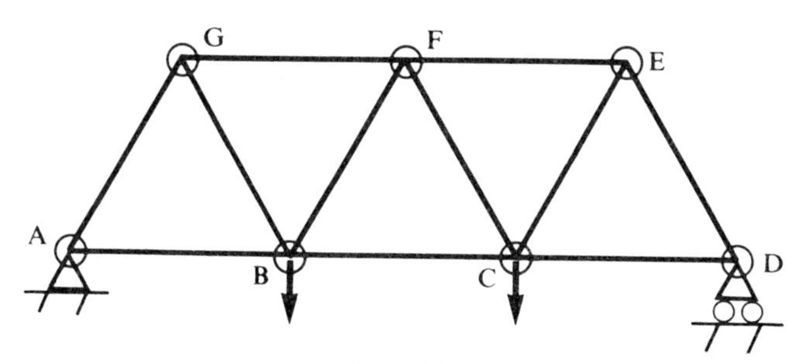

FIG. 6.15.

3. Consider again the pin-jointed framework previously dealt with in question 4 in Section 3.6 (page 48), and illustrated in Figure 3.23. Its joint coordinates are given in metres. Each of its bars consists of an aluminium tube, 100 mm in outside diameter and with a wall-thickness of 5 mm.

The framework carries a vertical load of 10 kN at D. At the relevant stress levels, aluminium obeys Hooke's law, with an elastic modulus E equal to $70\,\text{kN/mm}^2$. Estimate the three components of the deflection of D under the load.

4. Figure 6.16 shows a square pin-jointed framework pinned to a rigid abutment. Each of its bars has the same relation between axial tension and elongation, namely,

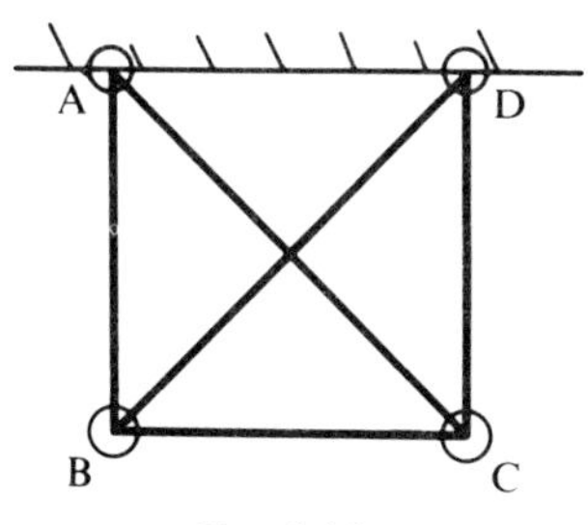

FIG. 6.16.

elongation/length = k(tension). Initially all the bars fit together perfectly. A vertical load P is then applied at B. Would you expect the tension induced in BC to be positive or negative? Find the tension in BC when P is 10 kN. A second vertical load Q is then applied at C. Find the tensions in BC when Q is 10 kN and P is zero, and when P and Q are both 10 kN.

5. Consider again the pin-jointed framework in Figure 6.17 but now let the elongation of each bar be related to its tension by

$$\text{elongation/length} = 10^{-7}T(1+cT^2),$$

where the tension T is measured in N. Sketch tension/elongation relations for $c>0$ and $c<0$. Show that the tension R in BC is given by

$$0=(P+R)\{1+c(P+R)^2\}+(Q+R)\{1+c(Q+R)^2\}+R(1+cR^2)$$
$$+4R\surd 2(1+2cR^2).$$

and find R when c is $10^{-8}\,\text{N}^{-2}$, P is 10 kN, and Q is zero. In Problem 4 you found the tension in BC when P and Q were both 10 kN to be twice the tension when P was 10 kN and Q was zero: would you expect that still to be true here?

6. In the pin-jointed plane framework shown in Figure 6.17 all the bars obey Hooke's law and have the same elastic modulus E. The cross-sectional area of all the bars except BD is a, and that of BD is $a/4$. The abutments at A and C are rigid. A vertical load W is applied at B. Show that the outward thrust induced at the abutments A and C is $0{\cdot}92\,W$, and find the vertical deflection of D.

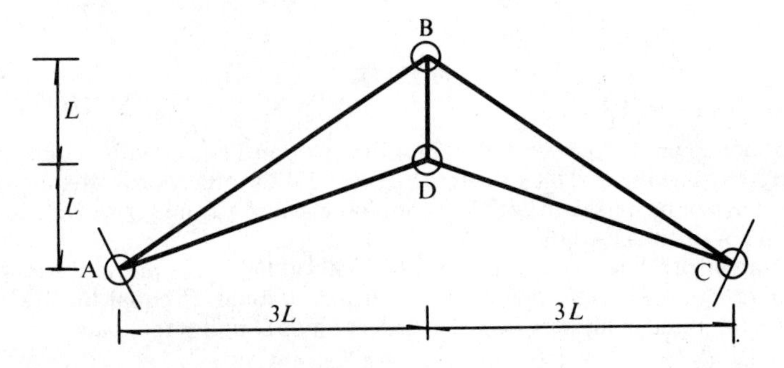

FIG. 6.17.

7. Rigid-jointed structures

7.1. Virtual work

VIRTUAL work is such a useful concept in structural mechanics that one naturally wishes to extend its application beyond pin-jointed frameworks, to other structures that have more complicated deformations and internal forces. In this section its extension to frame structures is examined. We first try to guess what the generalized statement of virtual work will turn out to be like, and then prove the statement as it applies to a restricted class of frames. A similar statement applies to any skeletal structure, and indeed to any solid body, but that general result will not be proved, because the proof is lengthy and demands mathematical knowledge that the reader of this book may not have.

Let us first of all go back once more to the result proved for pin-jointed frameworks,

$$\sum_{\text{all joints}} (U_i u_i + V_i v_i) = \sum_{\text{all bars}} T_{ij} e_{ij}. \qquad (6.2.23)$$

The external loads U_i and V_i are in equilibrium with internal tensions T_{ij}, and an independent set of displacements, u_i and v_i, are compatible with internal bar elongations e_{ij}. This suggests that the generalized statement will include a set of quantities describing external loads and internal forces in equilibrium (described through concepts like tension, bending moment, torque, stress, and so on), and an independent set of quantities describing a deformation which is geometrically compatible (described through concepts like displacement, elongation, curvature, strain, and so on). In the result for frameworks the external virtual work term sums the virtual work done by each load on its corresponding displacement. Thus, U_k is the horizontal load at K, and the corresponding displacement is u_k, the horizontal displacement at K, and so the product $U_k u_k$ contributes to the final sum for the virtual work. Each load and its corresponding displacement contribute to the sum over the whole structure. This suggests that we might conjecture that the correct generalization for the external virtual work is

$$\sum_{\text{all loads}} \text{(load) (its corresponding displacement)}.$$

Now consider the internal virtual work, corresponding to the sum over all bars of $T_{ij} e_{ij}$. Imagine first of all a straight bar, which elongates from an initial reference state in such a way that the amount of stretching varies along its length. Concentrate on one element, shown in Figure 7.1(a), initially ds

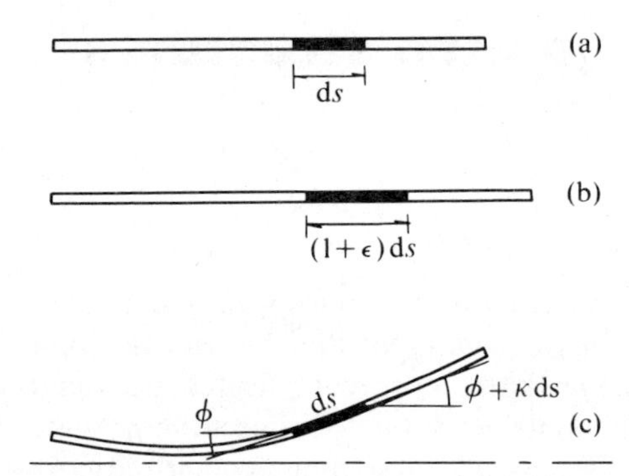

FIG. 7.1. (a) An element of a straight bar. (b) The same element, after a stretching deformation. (c) The same element, after a bending deformation.

long. After the deformation its length is $(1+\varepsilon)\,\mathrm{d}s$: it has lengthened by $\varepsilon\,\mathrm{d}s$, so that ε is the elongation per unit length at the element (Figure 7.1(b)). Since $\varepsilon\,\mathrm{d}s$ is the elongation of the element in this particular deformation, its contribution to the internal virtual work ought to be this elongation multiplied by the local corresponding internal force resultant, which is T, the tension at this point on the bar. The contribution is therefore $T\varepsilon\,\mathrm{d}s$; for the whole bar, it will be

$$\int_{\text{whole bar}} T\varepsilon\,\mathrm{d}s.$$

More tentatively, we can conjecture that this will still be the case even if the bar is curved.

This allows for stretching deformation. Elongation per unit length ε is associated with axial tension T, one of the stress resultants into which the resultant force and moment over a section of a skeletal structure can be divided. Other kinds of deformation will be associated with different stress resultants. Suppose that instead of stretching without bending the element in Figure 7.1 bends without stretching, so that its curvature alters from 0 to κ (Figure 7.1(c)). One end rotates through an angle ϕ, the other through $\phi+\kappa\,\mathrm{d}s$, recalling the definition of κ in Chapter 5. The work done on the element in such a deformation by a moment M acting on its ends will be

$$\underbrace{M(\phi+\kappa\,\mathrm{d}s)}_{\text{right-hand end}}-\underbrace{M\phi}_{\text{left-hand end}}$$

that is, $M\kappa\,\mathrm{d}s$. That suggests that we might conjecture that the internal virtual work associated with bending is $M\kappa\,\mathrm{d}s$ integrated over the whole structure, so that the internal virtual work is

$$\underbrace{\int T\varepsilon \, \mathrm{d}s}_{\text{stretching}} + \underbrace{\int M\kappa \, \mathrm{d}s}_{\text{bending}} + \text{(additional terms for other kinds of deformation)}.$$

All this has been conjecture. It is by no means a proof, although the result is in fact correct. It has been an exploratory argument, running ahead of proof, rather like a reconnaisance, a kind of argument that is often called heuristic (after Greek *heuriskō*: to discover). Such arguments are often useful, as long as we take care not to confuse them with proof.

We now prove the result, for a bar which lies in a plane and deforms in the same plane, and is subject to in-plane loading. It bends, but does not elongate: this restriction is made for simplicity, and could be dropped at the cost of lengthening the argument. As in the proof for plane frameworks, it is useful first to remind ourselves of a piece of elementary mathematics.

Integration by parts

Two quantities, f and g, are functions of a single variable s. Differentiate their product with respect to s

$$\frac{\mathrm{d}}{\mathrm{d}s}(fg) = f\frac{\mathrm{d}g}{\mathrm{d}s} + g\frac{\mathrm{d}f}{\mathrm{d}s}. \tag{7.1.1}$$

Rearranging,

$$f\frac{\mathrm{d}g}{\mathrm{d}s} = \frac{\mathrm{d}}{\mathrm{d}s}(fg) - g\frac{\mathrm{d}f}{\mathrm{d}s}. \tag{7.1.2}$$

Now integrate, with respect to s, from $s = a$ to $s = b$

$$\int_a^b f\frac{\mathrm{d}g}{\mathrm{d}s}\mathrm{d}s = [fg]_a^b - \int_a^b g\frac{\mathrm{d}f}{\mathrm{d}s}\mathrm{d}s, \tag{7.1.3}$$

where the expression

$$[fg]_a^b$$

signifies the difference between the value of fg when $s = b$ and the value when $s = a$. This process is called integration by parts, and is useful in the evaluation of integrals like the left-hand side of eqn (7.1.3).

Equilibrium of a curved bar

Consider the curved bar shown in Figure 7.2(a). It lies in a plane containing reference axes 1 (horizontal) and 2 (vertical). Position on the bar is defined by a distance s, measured along the bar from the end 0. The initial shape of the bar is defined by a function $\psi_0(s)$, which describes the angle ψ

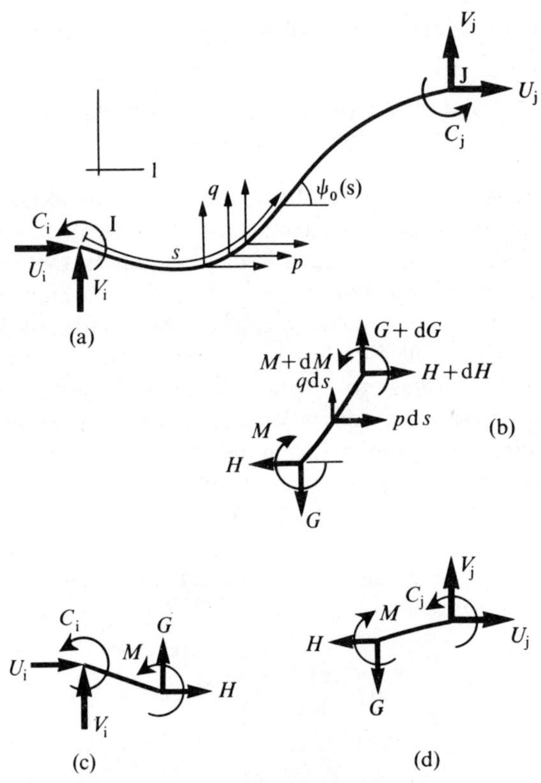

FIG. 7.2. (a) Forces and moments on a curved bar. (b) Forces and moments on an element of the bar. (c) Forces and moments on a short element at end I of the bar. (d) Forces and moments on a short element at end J of the bar.

between the bar axis and the 1-axis at a distance s from the end 0. Thus

$$\psi_0(s) = \alpha \quad \text{for all } s \tag{7.1.4}$$

would describe a straight bar making an angle α with the 1-axis, whereas

$$\psi_0(s) = \beta s \tag{7.1.5}$$

would describe an arc of a circle of radius $1/\beta$, tangential to the 1-axis at 0, and so on. There is no restriction on the initial shape, except that ψ is continuous, without jumps. The initial curvature of the bar is $d\psi_0/ds$.

The bar is loaded by end forces U_i and V_i at end i, U_j and V_j at end j, by end moments C_i and C_j (counter-clockwise positive), and by distributed loads along its length, p per unit length horizontally and q per unit length vertically. Consider an element ds (Figure 7.2(b)). The resultant force across a section is divided into horizontal and vertical components, and in addition there is a bending moment. At the end of the element closer to 0, their values are H, G, and M. In general they will not be constant, and at the other end of the element their values are $H+\mathrm{d}H$, $G+\mathrm{d}G$, and $M+\mathrm{d}M$. The external loads on the element are $p\,\mathrm{d}s$ and $q\,\mathrm{d}s$. The element must be in equilibrium, and the three equilibrium conditions give

$$\frac{\mathrm{d}H}{\mathrm{d}s}=-p \tag{7.1.6}$$

$$\frac{\mathrm{d}G}{\mathrm{d}s}=-q \tag{7.1.7}$$

$$\frac{\mathrm{d}M}{\mathrm{d}s}=H\sin\psi-G\cos\psi. \tag{7.1.8}$$

Deformation of a curved bar

If the bar bends without stretching (Figure 7.3(a)), its new configuration can be described in just the same way as the original shape, by the dependence of the angle ψ on the distance s from 0. Let

$$\psi(s)=\psi_0(s)+\phi(s). \tag{7.1.9}$$

If ϕ is zero for all s, there has been no deformation. The function $\phi(s)$ is a measure of how far the bar has deformed. The new curvature of the bar is $\mathrm{d}\psi/\mathrm{d}s$, and its change in curvature κ is $\mathrm{d}\phi/\mathrm{d}s$.

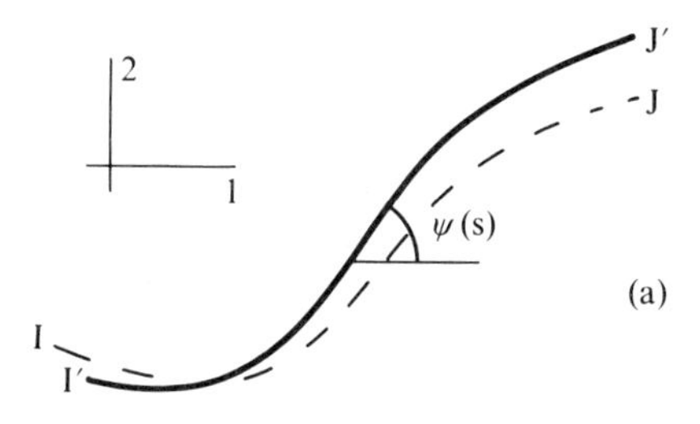

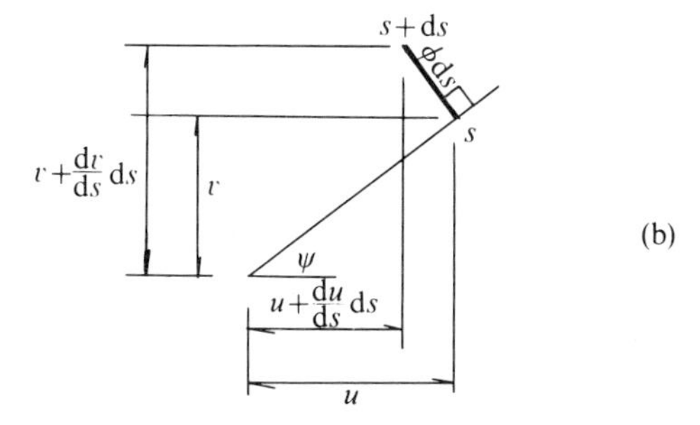

FIG. 7.3. (a) Deformation of a curved bar. (b) Displacement diagram for an element ds of a curved bar.

We now want to relate ϕ and ψ to displacements u and v, in the usual notation, referred to the 1- and 2-axes. If the bar bends without stretching, the elongation of ds is zero, and therefore any relative displacement of the ends of the element is due to rotation. The rotation of the element is ϕ, and so the relative displacement of the ends is ϕ ds. If ϕ is small—as we shall assume from now on—the displacement diagram for the element is Figure 7.3(b). The displacements of the left-hand end of the element are u and v, and those of the right-hand end are $u+(\mathrm{d}u/\mathrm{d}s)\,\mathrm{d}s$ and $v+(\mathrm{d}v/\mathrm{d}s)\,\mathrm{d}s$. From the geometry of the diagram

$$\frac{\mathrm{d}u}{\mathrm{d}s}=-\phi\sin\psi \tag{7.1.10}$$

$$\frac{\mathrm{d}v}{\mathrm{d}s}=\phi\cos\psi \tag{7.1.11}$$

Virtual work for a curved bar

A conjectural argument in the introduction to this section suggested that $M\kappa$ ds would be the virtual work associated with the bending of an element ds. Following this conjecture, we integrate $M\kappa$ over the whole bar, from I to J,

$$\int_{\mathrm{I}}^{\mathrm{J}} M\kappa\,\mathrm{d}s=\int_{\mathrm{I}}^{\mathrm{J}} M\frac{\mathrm{d}\phi}{\mathrm{d}s}\mathrm{d}s, \tag{7.1.12}$$

$$=[M\phi]_{\mathrm{I}}^{\mathrm{J}}-\int_{\mathrm{I}}^{\mathrm{J}}\phi\frac{\mathrm{d}M}{\mathrm{d}s}\mathrm{d}s, \tag{7.1.13}$$

by integration by parts, eqn (7.1.3),

$$=[M\phi]_{\mathrm{I}}^{\mathrm{J}}-\int_{\mathrm{I}}^{\mathrm{J}}\phi(H\sin\psi-G\cos\psi)\,\mathrm{d}s, \tag{7.1.14}$$

substituting from eqn (7.1.8),

$$=[M\phi]_{\mathrm{I}}^{\mathrm{J}}+\int_{\mathrm{I}}^{\mathrm{J}}\left(H\frac{\mathrm{d}u}{\mathrm{d}s}+G\frac{\mathrm{d}v}{\mathrm{d}s}\right)\mathrm{d}s, \qquad (7.1.15);$$

using eqns (7.1.10) and (7.1.11),

$$=[M\phi+Hu+Gv]_{\mathrm{I}}^{\mathrm{J}}-\int_{\mathrm{I}}^{\mathrm{J}}\left(u\frac{\mathrm{d}H}{\mathrm{d}s}+v\frac{\mathrm{d}G}{\mathrm{d}s}\right)\mathrm{d}s, \tag{7.1.16}$$

integrating by parts a second time,

$$=[M\phi + Hu + Gv]_{\mathrm{I}}^{\mathrm{J}} + \int_{\mathrm{I}}^{\mathrm{J}} (pu + qv)\,\mathrm{d}s, \qquad (7.1.17)$$

using eqns (7.1.6) and (7.1.7).

The final term in this equation is the work done by the distributed loads on their corresponding displacements, so that $p\,\mathrm{d}s$ is the horizontal load on element $\mathrm{d}s$, and u its horizontal displacement, and so on. This is as we should expect. The first term on the right is expressed in terms of the internal stress resultants H, G, and M, and they have to be related to the external loading at the ends. Figure 7.2(c) shows an infinitesimal element at end I; since it is in equilibrium

$$\left.\begin{aligned} H &= -U_{\mathrm{i}} \\ G &= -V_{\mathrm{i}} \\ M &= -C_{\mathrm{i}} \end{aligned}\right\} \text{at end I}, \qquad (7.1.18)$$

and therefore at end I

$$M\phi + Hu + Gv = -(U_{\mathrm{i}}u_{\mathrm{i}} + V_{\mathrm{i}}v_{\mathrm{i}} + C_{\mathrm{i}}\phi_{\mathrm{i}}). \qquad (7.1.19)$$

Figure 7.2(d) shows an infinitesimal element at end J; since it is in equilibrium

$$\left.\begin{aligned} H &= U_{\mathrm{j}} \\ G &= V_{\mathrm{j}} \\ M &= C_{\mathrm{j}} \end{aligned}\right\} \text{at end J}, \qquad (7.1.20)$$

and therefore at end J

$$M\phi + Hu + Gv = U_{\mathrm{j}}u_{\mathrm{j}} + V_{\mathrm{j}}v_{\mathrm{j}} + C_{\mathrm{j}}\phi_{\mathrm{j}}. \qquad (7.1.21)$$

Now substituting into eqn (7.1.17), which includes the difference between the values of $M\phi + Hu + Gv$ at the ends I and J, and rearranging,

$$\underbrace{(U_{\mathrm{i}}u_{\mathrm{i}} + V_{\mathrm{i}}v_{\mathrm{i}} + C_{\mathrm{i}}\phi_{\mathrm{i}})}_{\text{loads at end I}} + \underbrace{(V_{\mathrm{j}}u_{\mathrm{j}} + V_{\mathrm{j}}v_{\mathrm{j}} + C_{\mathrm{j}}\phi_{\mathrm{j}})}_{\text{loads at end J}} + \underbrace{\int_{\mathrm{I}}^{\mathrm{J}} (pu + qv)\,\mathrm{d}s}_{\text{loads between I and J}}$$

$$= \int_{\mathrm{I}}^{\mathrm{J}} M\kappa\,\mathrm{d}s. \qquad (7.1.22)$$

As we conjectured, the external virtual work includes contributions from all the external loads, each one multiplied by its own corresponding displacement. Notice again that the set of forces and moments in equilibrium is independent of the deformation. It has not been assumed that one is the cause of the other, and either may be chosen arbitrarily.

In this instance the deformation was inextensional, and so no elongation term enters the equation. It is not hard to generalize the argument, and then—as we guessed earlier—the result is

$$\sum_{\text{all loads}} (\text{load}^*)(\text{displacement}^{**}) = \int_{\substack{\text{whole}\\ \text{structure}}} M^*\kappa^{**}\,\mathrm{d}s + \int_{\substack{\text{whole}\\ \text{structure}}} T^*\varepsilon^{**}\,\mathrm{d}s + \begin{pmatrix}\text{terms for other kinds}\\ \text{of deformation}\end{pmatrix}. \tag{7.1.23}$$

In order to emphasize the independence of the equilibrium set of forces from the compatible deformation, the quantities which belong to the equilibrium set have been labelled* and those which belong to the compatible set have been labelled**.

The result can be generalized further to structures with more than one bar, and to three-dimensional structures. As we saw in Chapter 4, the internal forces are then more complicated. Table 7.1 lists the internal virtual work terms associated with torsion and multi-axis bending.

TABLE 7.1

Stress resultant	Associated measure of deformation	Contribution to virtual work
torque, Ω	ω, twist per unit length	$\int \Omega\omega\,\mathrm{d}s$
M_2 defined in Section 2.5	κ_2, change in curvature in local 1′, 3′ plane	$\int M_2\kappa_2\,\mathrm{d}s$
M_3 defined in Section 2.5	κ_3, change in curvature in local 1′, 2′ plane	$\int M_3\kappa_3\,\mathrm{d}s$

It is possible to generalize the whole idea still more, to two-dimensional surfaces and three-dimensional solid bodies. The general result for solid bodies has an attractive simplicity:

$$\text{virtual work of external loads} = \int_{\text{whole body}} (\text{stress})(\text{strain})\,\mathrm{d}V. \tag{7.1.24}$$

That, however, takes us beyond the scope of this book.

7.2. Deflections of beams and frames

The first application of the virtual work result derived in the preceding section will be to a beam deflection problem, analysed in Chapter 5 by direct integration of the governing differential equation.

Example 7.2.1 A cantilever of length L carries a uniformly distributed downward load of intensity w per unit length (Figure 7.4(a)). The bending moment is related to the curvature by

$$M = F\kappa,$$

where F is a constant flexural rigidity.

(i) *Find the deflection at the end of the cantilever.*

(ii) *Find the deflection at a point distant x from the left-hand end.*

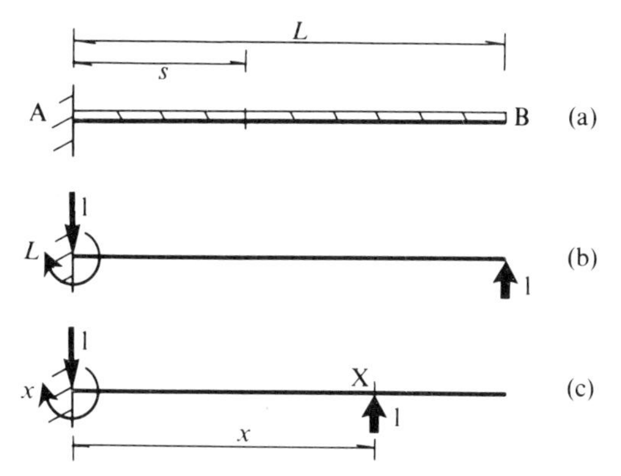

FIG. 7.4. (a) Uniformly loaded cantilever. (b) Unit load at the free end. (c) Unit load at a distance x from the built-in end.

The method is very like the one used to find the deflection of frameworks. Since we are interested in the actual deflection, we put into the virtual work statement the actual change in curvature κ and the actual deflections v. However, we choose as the equilibrium set of forces and moments one that corresponds to an imaginary unit load at the point whose deflection we wish to find.

It was shown earlier (Example 4.1.2, page 53), that at a point s from the left-hand end

$$M = -\tfrac{1}{2}w(L-s)^2. \tag{7.2.1}$$

The curvature κ is therefore

$$\kappa = -\tfrac{1}{2}w(L-s)^2/F. \tag{7.2.2}$$

In order to find the deflection at the end, imagine a unit load at one end

(Figure 7.4(b)). Its corresponding moment at a point located by s can be found in the usual way (by making an imaginary cut at S and considering the equilibrium of the segment to the right of it), and is

$$M^* = +(L-s). \tag{7.2.3}$$

The choice of unit load at the free end also implies a unit reaction at the built-in end, and an external moment L. Now choose

(i) the external loads and moments in the virtual work statement (7.1.23) to be those corresponding to the imaginary unit load (Figure 7.4(b)), and

(ii) the displacements and curvature to be those corresponding to the actual loading (Figure 7.4(a)).

Then

$$1 \,.\, v_b + (-1) \,.\, 0 + (L) \,.\, 0 = \int_{\substack{0 \\ \text{whole} \\ \text{beam}}}^{L} M^* \kappa \, ds \tag{7.2.4}$$

$$= \int_0^L (L-s)\{-\tfrac{1}{2}w(L-s)^2/F\}\, ds \tag{7.2.5}$$

$$v_b = -\left[\frac{w}{8F}(L-s)^4\right]_0^L \tag{7.2.6}$$

$$= -wL^4/8F. \tag{7.2.7}$$

The reactions at A contribute nothing to the external virtual work, not because they are themselves zero, but because in the actual displacement set there is no displacement at A and no rotation at A, by the definition of the problem.

If we require the displacement at a general point X, distant x from the left-hand end, the imaginary unit load should go at X. (Figure 7.4(c)). The corresponding moment at s from the left-hand end is

$$M^{**} = \begin{cases} x-s & 0<s<x \\ 0 & x<s<L. \end{cases} \tag{7.2.8}$$

Now choose these as the external loads and moments, but again choose the actual displacements and curvatures, and apply virtual work.

$$1 \,.\, v_x + (-1)\,.\,0 + (-x)\,.\,0 = \int_{\substack{\text{whole}\\ \text{beam}}} M^{**}\kappa \,\mathrm{d}s \tag{7.2.9}$$

$$= \int_0^x (x-s)\{-\tfrac{1}{2}w(L-s)^2/F\}\,\mathrm{d}s + \int_x^L 0\,.\{-\tfrac{1}{2}w(L-s)^2/2F\}\,\mathrm{d}s \tag{7.2.10}$$

integrating over the two segments separately, because of the different expressions for M^{**} on either side of X. The integration gives

$$v_x = -\frac{w}{2F}\left(\frac{x^4}{12} - \frac{x^3L}{3} + \frac{x^2L^2}{2}\right), \tag{7.2.11}$$

which agrees with the expression derived on page 74.

Direct integration of the governing equation cannot be used if beams are curved, but virtual work can be used.

Example 7.2.2. Figure 7.5 shows a curved bar, lying in a vertical plane, in the shape of a semi-circle, built-in at A and free at D. Its flexural rigidity is uniformly F. It deflects under a vertical load W applied at B. What are

(i) *the horizontal deflection at C,*

(ii) *the vertical deflection at B?*

Deflections due to axial extension of the bar are negligible.

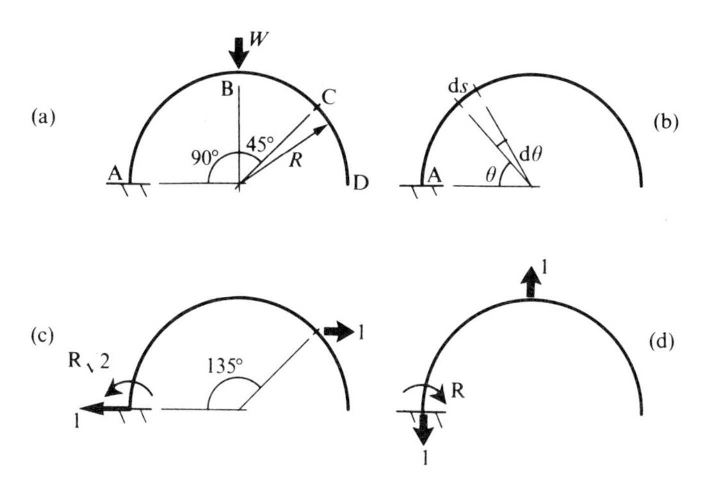

FIG. 7.5. (a) Curved bar, built-in at one end, and carrying a vertical load. (b) Location of point on bar by angle θ. (c) Unit horizontal load at C. (d) Unit vertical load at B.

It is simplest to locate points on the bar by an angle θ subtended at the centre of the semi-circle (Figure 7.5(b)). If s is measured along the bar from A

$$s = R\theta \tag{7.2.12}$$

and a length element is

$$ds = R\,d\theta. \tag{7.2.13}$$

Once again, since we are concerned with actual deflections, we put actual curvatures κ and deflections u_c, v_c, u_b, and so on into the virtual work equation. At a point S, between A and B, the bending moment is

$$M = -WR\cos\theta, \tag{7.2.14}$$

but to the right of B the bending moment is zero. Accordingly

$$\kappa = \begin{cases} -(WR/F)\cos\theta & 0<\theta<\pi/2 \\ 0 & \pi/2<\theta<\pi. \end{cases} \tag{7.2.15}$$

Following the same line of attack as in the last example, we find the horizontal deflection at C by thinking of an imaginary unit horizontal load at C (Figure 7.5). Call the corresponding bending moment $M^{(1)}$. Then

$$M^{(1)} = \begin{cases} +(\sin\theta - 1/\sqrt{2})R, & 0<\theta<3\pi/4, \\ 0, & 3\pi/4<\theta<\pi. \end{cases} \tag{7.2.16}$$

Applying virtual work, using these forces and moments and the actual curvatures

$$1\,.\,u_c + (-1)\,.\,0 + \frac{R}{\sqrt{2}}\,.\,0 = \int_{\text{whole bar}} M^{(1)}\kappa\,ds, \tag{7.2.17}$$

$$= \int_0^{\pi/2} (\sin\theta - 1/\sqrt{2})R\left\{-\frac{WR}{F}\cos\theta\right\}R\,d\theta, \tag{7.2.18}$$

the integral for θ greater than $\pi/2$ being zero because the integrand is then zero, and so

$$u_c = \left(\frac{1}{\sqrt{2}} - \frac{1}{2}\right)WR^3/F. \tag{7.2.19}$$

In order to find the vertical deflection at B, we need an imaginary unit vertical load at B (Figure 7.5(d)). Call the corresponding bending moment $M^{(2)}$;

$$M^{(2)} = \begin{cases} R\cos\theta, & 0<\theta<\pi/2, \\ 0, & \pi/2<\theta<\pi. \end{cases} \tag{7.2.20}$$

Applying virtual work, still with the actual curvatures, but now with this second system of forces and bending moments,

$$1 \,.\, v_b + (-1)\,.\,0 + (-R)\,.\,0 = \int_{\text{whole bar}} M^{(2)}\kappa \, ds \tag{7.2.21}$$

$$= \int_0^{\pi/2} (R \cos\theta)\left\{-\frac{WR}{F}\cos\theta\right\} R \, d\theta \tag{7.2.22}$$

$$= -\pi WR^3/4F. \tag{7.2.23}$$

It happens that in both these examples the change in curvature has been linearly proportional to the bending moment. The method can still be used if this is not so; see problem 2 in Section 7.7.

7.3. Statically indeterminate beams and frames

The forces within statically indeterminate beams and frames can be found by a method essentially identical with the one used to find tensions in the bars of statically indeterminate frameworks (Section 6.4).

Example 7.3.1. A straight continuous beam rests on three simple supports, A, C, and D (Figure 7.6). Its bending moment is related to its curvature by a uniform flexural rigidity F. Initially, before the beam is loaded, the support reactions are all zero. A uniform distributed load w per unit length is then applied to the beam. What are

(i) *the reactions at the supports,*
(ii) *the deflection at the centre of the left-hand span?*

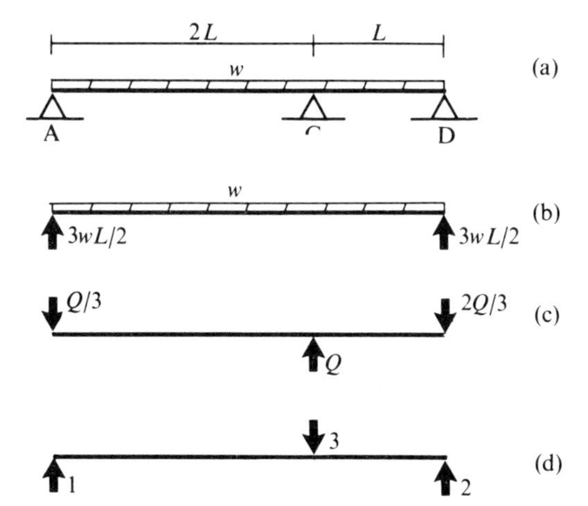

FIG. 7.6. (a) Two-span continuous beam. (b) Loading and end reactions if reaction Q at centre support is zero. (c) Loading and end reactions if external load is absent, but centre support reaction Q is not zero. (d) Self-equilibrating system of end reactions corresponding to $Q = -3$.

This structure is once redundant: if one of the supports were removed it would be statically determinate. Alternatively: there are three unknown vertical reactions at the supports, but only two equilibrium equations to determine them.

Call the middle support reaction Q. As when we dealt with pin-jointed frameworks, think of the force system as the sum of two systems, the first one with the external loads but with the redundant force absent, the second with no external loads but with the redundant force present. The bending moment corresponding to the first of these (Figure 7.6(b)) is

$$M=\tfrac{1}{2}wx(3L-x), \tag{7.3.1}$$

measuring x from the left-hand support, and that corresponding to the second (Figure 7.6(c)) is

$$M=\begin{cases}-Qx/3 & x<2L,\\ -2Q(3L-x)/3 & x>2L.\end{cases} \tag{7.3.2}$$

Combining these to find the actual curvature,

$$\kappa=\begin{cases}\{\tfrac{1}{2}wx(3L-x)-Qx/3\}/F, & x<2L,\\ \{\tfrac{1}{2}wx(3L-x)-2Q(3L-x)/3\}F, & x>2L.\end{cases} \tag{7.3.3}$$

Only bending deformation occurs, and all the loads are vertical. The virtual work statement, eqn (7.1.23), then takes the form

$$\sum_{\substack{\text{all}\\ \text{concentrated}\\ \text{loads}}} V_i^* v_i^{**} + \int_{\substack{\text{distributed}\\ \text{load}}} q^* v^{**}\,\mathrm{d}s = \int_{\substack{\text{whole}\\ \text{structure}}} M^*\kappa^{**}\,\mathrm{d}s. \tag{7.3.4}$$

Identify the compatible deformation, marked **, with the actual curvature κ and the corresponding displacements v. Choose the equilibrium system of forces, marked *, so that there are no external loads except at the supports. Then

$$\sum V_i^* v_i^{**}=0, \tag{7.3.5}$$

because there are no external loads except at the supports, and there v is zero (because the supports do not move), and

$$\int q^* v^{**}\,\mathrm{d}s=0 \tag{7.3.6}$$

because q^* is zero everywhere, and so

$$0=\int_{\substack{\text{whole}\\ \text{structure}}} M^*\kappa\,\mathrm{d}s. \tag{7.3.7}$$

Already we have a system of forces which meets these special requirements, if in the second distribution mentioned above (eqn (7.3.2), Figure 7.6(c)) we replace Q by (say) -3, which makes the arithmetic simpler, so that the force distribution is that in Figure 7.6(d) and

$$M^* = \begin{cases} x, & x < 2L, \\ 6L - 2x, & x > 2L. \end{cases} \tag{7.3.8}$$

Substituting into eqn (7.3.7), for κ from eqn (7.3.3) and for M^* from eqn (7.3.8),

$$0 = \int_0^{2L} x\{\tfrac{1}{2}wx(3L-x) - Qx/3\}/F\,\mathrm{d}x + \int_{2L}^{3L} (6L-2x) \times\{\tfrac{1}{2}wx(3L-x) - 2Q(3L-x)/3\}/F\,\mathrm{d}x \tag{7.3.9}$$

$$= 11wL^4/4F - 4QL^3/3F \tag{7.3.10}$$

and so

$$Q = 33wL/16. \tag{7.3.11}$$

The other support reactions can now be calculated from statics. Figure 7.7 shows the calculated support reactions and the shear force and bending moment diagrams.

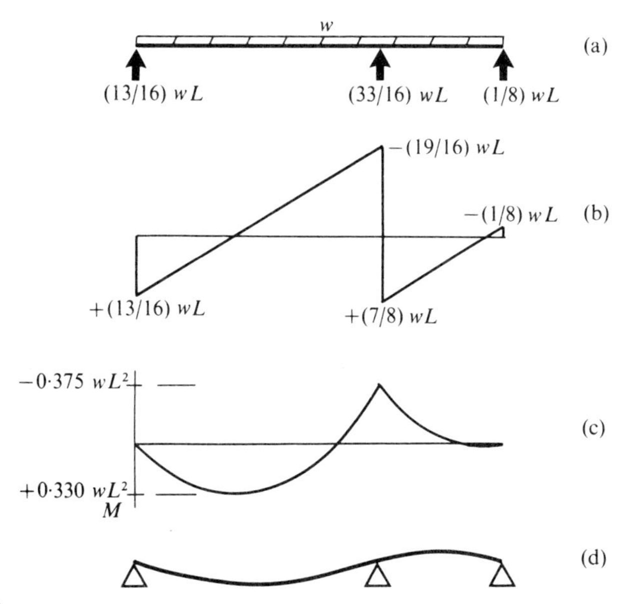

FIG. 7.7. (a) Calculated support reactions for Example 7.3.1. (b) Shear force diagram. (c) Bending moment diagram. (d) Distribution of deflection.

It is not wise to solve problems like this purely mathematically, without pausing to try to see what the results mean, and if they are physically sensible. Most people find it easier to see what deformations can be expected than to visualize the forces within a structure. Think of a beam resting on three supports, and then imagine it uniformly pressed down by vertical loads. If you find it difficult to see what will happen, make a simple model: rest a strip of cardboard on three pencils, and load it with coins or paper clips. It will take up the form shown in an exaggerated way in Figure 7.7(d). Most of the left-hand span sags downward, but over the central support the beam bends the other way, so that it is convex upward; structural engineers call this 'hogging'. A little to the left of the central support there is an inflection point, at which the curvature is zero. To the right of the inflection the curvature is hogging, and to the left sagging. Most of the right-hand span is bent upward. Comparison with Figure 7.7(c) shows that this is qualitatively consistent with the analysis: the signs of the curvatures are correct, the inflection point where the moment is zero is about where it would be expected, and so on. A discrepancy would not necessarily mean that the analysis was wrong, but it would indicate that a check would be wise.

Notice that the values of the support reactions do not depend on the absolute value of the flexural rigidty F, but that they do depend on the relative stiffness if different parts of the beam. If the left-hand span had instead a stiffness βF, the curvature in eqn (7.3.3) would be divided by β in the left-hand span, and this change would carry through into the first integral in eqn (7.3.9), and would alter the result.

Having found the bending moment at each point, we know the curvature everywhere, and the deflections can be found in the usual way, from the moments in equilibrium with an imaginary unit load, as in Examples 7.2.1 and 7.2.2. In a statically indeterminate structure, there is more than one possible distribution of bending moment in equilibrium with unit load, but any one will do. Thus, to find the central deflection v_b at the centre of the left-hand span, apply an imaginary unit load at that point. Figure 7.8 shows two of the infinitely many different distributions of support reactions which can be in equilibrium with this unit load. The distribution in Figure 7.8(a)

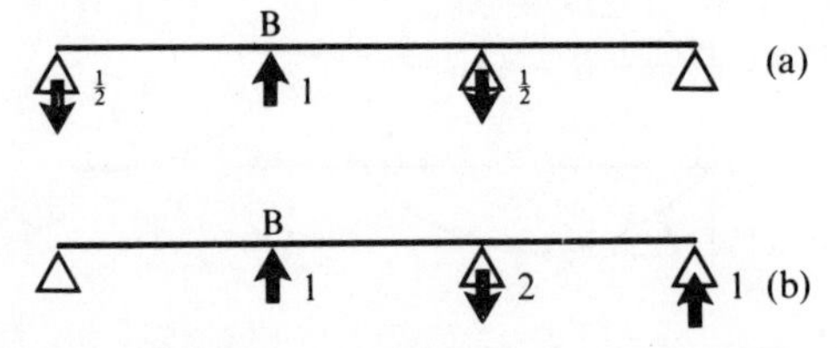

FIG. 7.8. (a) Distribution of support reactions in equilibrium with unit load at centre of left-hand span of continuous beam in Example 7.3.1. (b) Alternative distribution of support reactions in equilibrium with unit load at centre of left-hand span of continuous beam in Example 7.3.1.

gives

$$M^{(1)} = \begin{cases} -\tfrac{1}{2}x & x<L \\ -\tfrac{1}{2}(2L-x) & L<x<2L \\ 0 & x>2L, \end{cases} \tag{7.3.12}$$

whereas that in Figure 7.8(b) gives

$$M^{(2)} = \begin{cases} 0 & x<L \\ x-L & L<x<2L \\ 3L-x & x>2L. \end{cases} \tag{7.3.13}$$

Either of these distributions, which are different but correspond to the same unit load, gives the same deflection at the unit load point. The reason for this is that the differences between them are self-equilibrating, needing only reactions at the fixed support points to balance them, and we have already seen that any distribution of bending moment that is self-equilibrating satisfies eqn (7.3.7).

Since it does not matter which distribution of bending moment in equilibrium with unit load is used in the calculation of v_b, we choose the simplest, $M^{(1)}$, described by eqn (7.3.12). In the virtual work statement, eqn (7.3.4), we now make this the equilibrium set and the actual curvatures and displacements the compatible set. Then

$$\underset{\substack{\text{support}\\ \text{A}}}{(-\tfrac{1}{2}).0} + \underset{\text{B}}{1.v_b} + \underset{\substack{\text{support}\\ \text{C}}}{(-\tfrac{1}{2}).0} + \underset{\substack{\text{support}\\ \text{D}}}{0.0} = \int_{\substack{\text{whole}\\ \text{beam}}} M^{(1)}\kappa \,\mathrm{d}s. \tag{7.3.14}$$

Substituting from eqn (7.3.11) into eqn (7.3.3) to find κ, and from (7.3.12) for $M^{(1)}$,

$$v_b = \int_0^L (-\tfrac{1}{2}x)\{\tfrac{1}{2}wx(3L-x) - 11wLx/16\}/F\,\mathrm{d}x$$

$$+\int_L^{2L} -\tfrac{1}{2}(2L-x)\{\tfrac{1}{2}wx(3L-x) - 11wLx/16\}/F\,\mathrm{d}x \tag{7.3.15}$$

$$= -(11/96)wL^4/F \tag{7.3.16}$$

Example 7.3.2. A rectangular rigid-jointed portal frame is hinged to a rigid foundation (Figure 7.9(a)). It has a linear relation between bending moment and curvature, but the flexural rigidity is not constant throughout. The flexural rigidity of the beam is F, and that of the stanchions is αF; α is a constant. The frame carries a central load W. What bending moments are induced in it? Assume that only bending deformation occurs; the effects of axial deformation will be examined later.

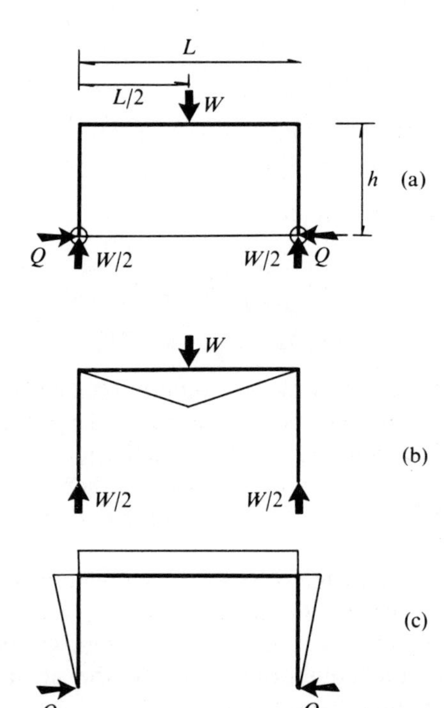

FIG. 7.9. (a) Symmetrical portal frame with hinged stanchion feet. (b) Support reactions if a concentrated load W acts at the centre of the beam, but the horizontal reaction at the stanchion feet is zero. (c) Support reactions if the horizontal reaction Q at the stanchion feet is non-zero, but there is no external load.

The hinges can rotate freely, so that the hinge moment is zero, but they do exert a horizontal reaction, preventing the feet of the stanchions from moving apart. Since the frame and the loading are symmetrical left to right, the vertical reaction at each stanchion foot is $\frac{1}{2}W$. The horizontal reactions must be equal and opposite; call them Q. Statics alone does not determine Q, and the frame is once redundant (once only, because when Q is found the bending moment and shear force are determined throughout the frame).

We proceed exactly as in the beam problem. The bending moment is the sum of

(1) the moments that would exist if W were present but Q were zero (Figure 7.9(b)), in which case

$$M = 0 \text{ in the stanchions}$$

$$M = \begin{cases} Wx/2 & x < \frac{1}{2}L \\ W(L-x)/2 & x > \frac{1}{2}L \end{cases} \quad \text{in the beam,} \qquad (7.3.17)$$

where x is measured along the beam from its left-hand end, and

(2) the moments that would exist if there were no load but Q were present (Figure 7.9(c)), in which case

$$M = -Qy \text{ in the stanchions,} \qquad (7.3.18)$$

at a distance y above the hinge, and

$$M = -Qh \text{ in the beam.} \qquad (7.3.19)$$

Figures 7.9(b) and (c) include superimposed bending moment diagrams, drawn in a convention which is often used in structural mechanics. The moment at each point is plotted at right angles to the frame; the direction of plotting is chosen so that if the diagram lies to the left of the frame, the corresponding moment would tend to make the frame bend so that it is convex to the left, if the diagram is below the frame, it would tend to bend so that it is convex downward (sagging), and so on.

Adding the two distributions of bending moment together

$$M = \begin{cases} -Qy & \text{in the stanchion,} \\ -Qh + Wx/2 & \text{in the left-hand half of the beam,} \\ -Qh + W(L-x)/2 & \text{in the right-hand half of the beam.} \end{cases} \qquad (7.3.20)$$

The curvature is the moment divided by the appropriate flexural rigidity,

$$\kappa = \begin{cases} -Qy/\alpha F & \text{in the stanchion, } 0 < y < h, \\ \{-Qh + Wx/2\}/F & \text{in the beam, } 0 < x < L/2, \\ \{-Qh + W(L-x)/2\}F & \text{in the beam, } L/2 < x < L. \end{cases} \qquad (7.3.21)$$

These are the actual curvatures of the frame. They must be compatible with zero displacements at the feet of the stanchions, so that after it has been deformed under the loading, Q takes such a value that the frame still 'fits' the foundation.

The argument is the same as in Example 7.3.1. If M^* is a distribution of bending moment in equilibrium with zero external loads, except at points which in the actual structure are known not to move, and we choose this as the equilibrium set in the virtual work statement, eqn (7.1.23), and the actual curvatures and displacements as the compatible set, then again

$$0 = \int_{\substack{\text{whole}\\ \text{frame}}} M^* \kappa \, ds. \qquad (7.3.22)$$

System (2) above has the properties we are looking for. If for simplicity we replace Q in eqn (7.3.18) and (7.3.19) by 1, we have

$$M^* = \begin{cases} -y & \text{in the stanchions,} \\ -h & \text{in the beam,} \end{cases} \qquad (7.3.23)$$

which corresponds to unit inward forces at the stanchion feet, and no other external loads.

The integral in eqn (7.3.22) is over the whole frame. However, since the frame is symmetric, and the integral over the whole frame is zero, so will be the integral over the left-hand half of the frame, and therefore, substituting for M^* from eqn (7.3.23) and for κ from eqn (7.3.21),

$$0 = \int_0^h (-y)\left\{-\frac{Qy}{\alpha F}\right\} \mathrm{d}y + \int_0^{\frac{1}{2}L} (-h)\{(-Qh + Wx/2)/F\}\, \mathrm{d}x \tag{7.3.24}$$

$$= Qh^3/3\alpha F - WhL^2/16F + Qh^2L/2F \tag{7.3.25}$$

and so

$$Q = \frac{WL}{8h(1 + 2h/3\alpha L)}. \tag{7.3.26}$$

Suppose, for instance, that

$W = 50$ kN

$L = 10$ m

$h = 5$ m

$\alpha = 1$ (beam and stanchions of same rigidity).

Then

$$Q = 9{\cdot}38 \text{ kN},$$

and the bending moment at the corner of the frame is

$$-Qh = -46{\cdot}9 \text{ kN m},$$

and that at the centre of the beam, under the load, is

$$-Qh + WL/4 = 78{\cdot}1 \text{ kN m}.$$

The bending moment diagram and the load on the frame are shown in Figure 7.10(a).

Is this the kind of deformation we expect? If we made such a frame, but put the lower ends of the stanchions on rollers instead of hinging them to a rigid foundation, and then applied a downward load to the beam, the stanchion feet would move outward, in the way shown in Figure 7.10(b). This is because the beam would sag downward, the left-hand and right-hand ends of the beam would rotate, clockwise and counter-clockwise respectively, and because the corner joints are rigid, the left and right stanchions would rotate about their upper ends, in the same directions. This can be observed in a simple model of wire or cardboard. In the case we analysed,

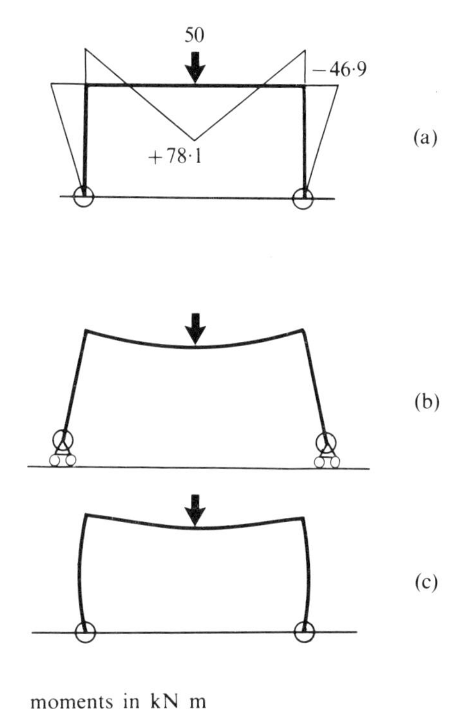

FIG. 7.10. (a) Distribution of bending moment in portal frame of Example 7.3.2. (b) Deflections that would occur if the stanchion feet were free to move sideways. (c) Deflections that occur if the stanchion feet are restrained.

however, the frame is pinned to a foundation which prevents the stanchion feet from moving apart. It must do this by exerting forces acting inwards on the feet, and so we would expect Q to be positive, which is how it came out in the calculation. Since the stanchions have to bend, moments are exerted on them by the beam. The sense of these moments is such that the ends of the beam are in hogging curvature, and the centre sags, so that there are inflection points some way in from the ends of the beam. This too agrees with the results given in Figure 7.10(a).

Example 7.3.3. A rectangular rigid-jointed portal frame has its stanchion feet built-in to a rigid foundation (Figure 7.11). It has a linear relation between bending moment and curvature, but the flexural rigidity is not constant. The flexural rigidity of the beam is F, and that of the stanchions is αF. The frame carries a central load W. What bending moments are induced in it?

This is Example 7.3.2 modified by the fixing of the stanchion feet, so that they cannot rotate. The frame and its loading are still symmetric about a central vertical axis, and so the vertical reactions at the stanchion feet are

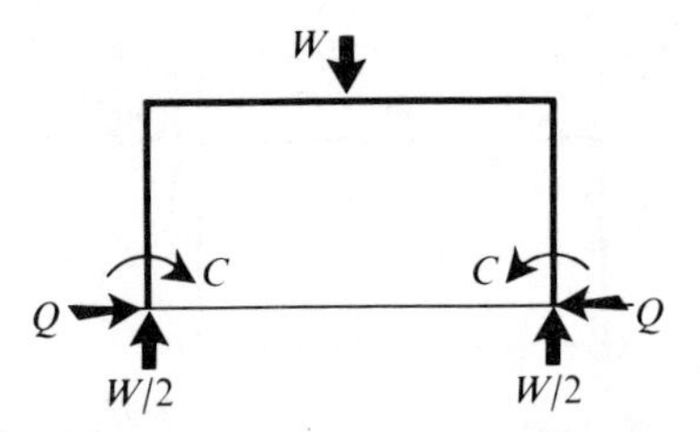

FIG. 7.11. Symmetrical portal frame with built-in stanchion feet.

still $\frac{1}{2}W$. At the stanchion feet act an unknown horizontal reaction Q and an unknown fixing moment C; symmetry tells us that the forces and moments at the left-hand and right-hand feet must be equal and opposite.

The method of solution follows Example 7.3.2, except that there are now two redundants, and so we now need two systems of moments in equilibrium with zero external loads except at the stanchion feet. The bending moment is

$$M = \begin{cases} C - Qy & \text{in the stanchion,} \\ C - Qh + Wx/2 & \text{in the left-hand half of the beam,} \end{cases} \tag{7.3.27}$$

using the notation of Example 7.3.2, and so the curvature is

$$\kappa = \begin{cases} (C - Qy)/\alpha F & \text{in the stanchion,} \\ (C - Qh + Wx/2)/F & \text{in the left-hand half of the beam.} \end{cases} \tag{7.3.28}$$

One suitable set of moments corresponds to setting Q equal to unity and C and W equal to zero in eqn (7.3.27). It is the set used in the previous example, and gives

$$M^{(1)} = \begin{cases} -y & \text{in the stanchions,} \\ -h & \text{in the beam.} \end{cases} \tag{7.3.29}$$

Choosing these moments (and the corresponding external loads), as the equilibrium set in the virtual work statement, and choosing the actual curvature κ and displacements as the compatible set,

$$0 = \int_{\substack{\text{whole}\\ \text{frame}}} M^{(1)} \kappa \, \mathrm{d}s. \tag{7.3.30}$$

A second set of moments $M^{(2)}$ corresponds to setting C equal to unity and Q and W to zero in eqn (7.3.27), so that

$$M^{(2)} = 1 \qquad \text{throughout the frame,} \tag{7.3.31}$$

which is equilibrated by unit moments at the stanchion feet. Making these the equilibrium moments

$$0 = \int_{\substack{\text{whole}\\ \text{frame}}} M^{(2)} \kappa \, \mathrm{d}s. \tag{7.3.32}$$

Equations (7.3.30) and (7.3.32) are independent simultaneous equations for C and Q, and can therefore be solved, and the distribution of bending moment throughout the frame can then be calculated.

Axial elongations in frames

In Example 7.3.2 above the stanchions carry compressive forces $\frac{1}{2}W$, and the beam carries a compressive force Q. Since the members of the frame carry axial forces, they will change in length. In the calculation, however, it was tacitly assumed that only bending deformation occurs: the internal virtue work included $\int M\kappa \, \mathrm{d}s$ but left out $\int T\varepsilon \, \mathrm{d}s$. Does this assumption have any justification?

This problem will be tackled by carrying out a reanalysis of Example 7.3.2, and showing that the effects of axial deformations are almost always very small. Return to Example 7.3.2, and suppose that the relation between tension T and elongation per unit length ε is linear (like that between bending moment and curvature), and that

$$\varepsilon = T/\Gamma, \tag{7.3.33}$$

where Γ is an axial stiffness, uniform over the whole frame, with the dimensions of force. The curvature is given by the same expression as before, eqn (7.3.21). In the stanchions the tension T is $-\frac{1}{2}W$, and so

$$\varepsilon = -W/2\Gamma, \tag{7.3.34}$$

while in the beam the tension is $-Q$, and so

$$\varepsilon = -Q/\Gamma. \tag{7.3.35}$$

The system of forces, moments M^*, and tensions T^* shown in Figure 7.9(c) (with Q replaced by 1), equilibrated only by reactions at the stanchion feet, has

$$M^* = \begin{cases} -y & \text{stanchions,} \\ -h & \text{beam,} \end{cases} \tag{7.3.23}$$

$$T^* = \begin{cases} 0 & \text{stanchions,} \\ -1 & \text{beam.} \end{cases} \tag{7.3.36}$$

By the usual argument

$$0 = \int_{\substack{\text{whole}\\\text{structure}}} M^*\kappa \, ds + \int_{\substack{\text{whole}\\\text{structure}}} T^*\varepsilon \, ds. \qquad (7.3.37)$$

The integrals can again be taken over the left-hand half of the frame, because of symmetry, and eqn (7.3.37) then gives

$$0 = Qh^3/3\alpha F - WhL^2/16F + Qh^2L/2F + QL/2\Gamma, \qquad (7.3.38)$$

the first three terms coming from the first integral in eqn (7.3.37), and therefore being exactly as in eqn (7.3.25). Therefore

$$Q = \frac{WL}{8h(1 + 2h/3\alpha L + F/\Gamma h^2)}. \qquad (7.3.39)$$

When we compare this with the solution obtained when axial deformation was left out (eqn (7.3.26), page 130), we see that the only change is the addition of $F/\Gamma h^2$ to the denominator. This term is normally very small: for instance, in a steel frame with h equal to 4 m, made from a 310 mm by 125 mm 48 kg/m universal beam section, bent about its major axis

$$F = 1{\cdot}99 \times 10^4 \text{ kN m}^2$$

$$\Gamma = 1{\cdot}28 \times 10^6 \text{ kN}$$

and

$$F/\Gamma h^2 = 10^{-3}.$$

The $F/\Gamma h^2$ term would only become important if h were very small, in fact so small that the height of the frame is of the same order as the depth of the section from which it is made. This conclusion can be generalized, and it turns out that axial deformation can be important in very flat arches, where the rise is of the same order as the thickness of the arch rib, but not in other frames or arches.

7.4. Statically indeterminate structures: effects of temperature changes

Imagine an unloaded statically determinate framework. Suppose some of the bars to alter slightly in length, perhaps as a result of temperature changes. Since the framework is statically determinate, these changes in length will not induce forces in the bars. The resulting deflections of the framework can be found by statics, in the usual way.

Now think of an unloaded indeterminate structure, which fits together exactly, without any forces in the bars, for instance the framework in Figure 7.12, which we have analysed before. Imagine AD to increase in length

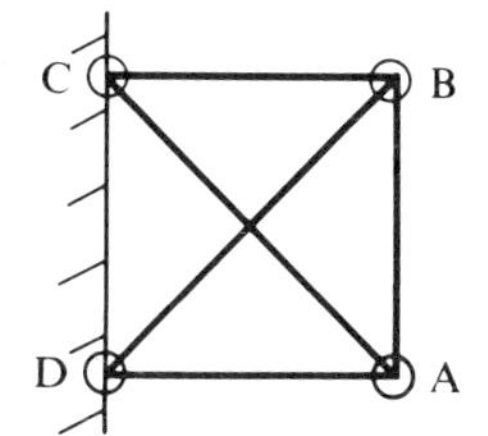

FIG. 7.12. Statically indeterminate framework.

slightly. It cannot do this without increasing the distance between A and D. However, relative motion between A and D cannot occur without elongations in at least some of the other bars, and these elongations can only develop as a result of forces set up in the bars. It follows that AD cannot expand freely: any change in its length will induce forces in the whole framework.

Another way of looking at it is this. Think of starting with the rigid foundation CD and building the framework bar by bar, leaving AD to the last. Bars BC and BD fix B, and then AB and AC fix A. At this stage the framework is still determinate, and all the tensions in the bars must still be zero. Suppose AD is a little longer than it ought to be to fit into the gap, either because its temperature has changed or because it was made wrongly. Then it can only be got into place by forcing A and D apart. It will itself be in compression, and forces will be set up in the rest of the structure.

The forces which are induced by this kind of effect can be found by virtual work. The only difficulty is that signs can be confusing if the problem is not solved systematically.

Think first of a reference configuration of a 'perfect' unloaded structure in which all the bars fit together precisely, and all the bar tensions are zero. In the real structure there are in general external loads, there are temperature changes, and some of the bars did not originally fit exactly. The actual configuration is a different one from the reference configuration. The set of elongations and deflections in going from the reference to the actual configuration must be compatible: this is merely a question of geometry, and it does not matter that the structure never actually occupies the reference configuration. The elongation from the reference configuration for each bar is the combined effect of three causes:

(i) tension in the bar,
(ii) temperature change,
(iii) initial lack of fit, so that even if there were no tension and no temperature change the bar would not match the reference configuration.

The total elongations e_{ij} are compatible. If T^*_{ij} is a set of bar tensions in equilibrium with zero external loads, except possibly at points constrained

TABLE 7.2

Bar	Length	Area	Tension T_{ij}	Elongation e_{ij}: Due to tension	Elongation e_{ij}: Due to lack of fit	Elongation e_{ij}: Due to temperature	T^*_{ij}	$T^*_{ij} e_{ij}$
BC	L	A	R	RL/AE	0	$\alpha\theta L$	1	$RL/AE + \alpha\theta L$
AB	L	A	R	RL/AE	0	0	1	RL/AE
AD	L	A	$-(P-R)$	$-(P-R)L/AE$	$-\Delta$	0	1	$-(P-R)L/AE - \Delta$
AC	$L\sqrt{2}$	$\frac{1}{2}A$	$(P-R)\sqrt{2}$	$4(P-R)L/AE$	0	$\alpha\theta L\sqrt{2}$	$-\sqrt{2}$	$-4\sqrt{2}(P-R)L/AE - 2\alpha\theta L$
BD	$L\sqrt{2}$	$\frac{1}{2}A$	$-R\sqrt{2}$	$-4RL/AE$	0	$\alpha\theta L\sqrt{2}$	$-\sqrt{2}$	$4RL\sqrt{2}/AE - 2\alpha\theta L$
								$(3+8\sqrt{2})RL/AE - (1+4\sqrt{2})PL/AE - 3\alpha\theta L - \Delta$

not to deflect, then, applying virtual work in the usual way

$$0 = \sum_{\text{all bars}} T^*_{ij} e_{ij}. \tag{7.4.1}$$

Example 7.4.1. A square pin-jointed framework ABCD (Figure 7.8) is attached to a rigid foundation at C and D. All the bars obey Hooke's law, and have the same elastic modulus E and linear thermal expansion coefficient α. Bars AB, BC, and DA have cross-sectional area A, and bars AC and BD have cross-sectional area $\frac{1}{2}A$. When the structure was made, at a uniform temperature θ_0, bar AD was Δ too short. Since then bars AC, BC, and BD have risen in temperature to $\theta_0 + \theta$, but the other bars have remained at θ_0. A vertical load P has been applied at A. What are the tensions in the bars?

A framework with the same geometry was analysed in Example 6.4.1, but there all the bars had the correct initial lengths and there were no temperature changes. If, as before, we call the tension in bar AB R, the tensions in the other bars are those shown in Figure 6.11(c), and listed in the fourth column of Table 7.2. The different components of the total elongation e_{ij} are listed in the fifth, sixth, and seventh columns. The self-equilibrating tensions T^*_{ij} are those in Figure 6.12, again as before, and are listed in the eighth column. The last column contains the different terms $T^*_{ij} e_{ij}$ in the summation of eqn (7.4.1). It follows that

$$0 = (3 + 8\sqrt{2})RL/AE - (1 + 4\sqrt{2})PL/AE - 3\alpha\theta L - \Delta, \tag{7.4.2}$$

and so

$$R = \frac{(1 + 4\sqrt{2})P + 3\alpha\theta AE + \Delta AE/L}{3 + 8\sqrt{2}} \tag{7.4.3}$$

and the other bar tensions can be calculated by substitution back into column four of Table 7.2.

Similar effects occur in rigid-jointed frames, and can be analysed in a similar way; see Example 7.9.

7.5. Degree of indeterminacy of frame structures

How do we know how many times redundant a frame is? It is not possible to answer this question as systematically as it is in the case of pin-jointed frameworks, but the same ideas can be used. One counts the unknown forces that would be needed completely to define the stress resultants within the structure, and then counts the equilibrium conditions that can be used to find them.

Consider, for example, the continuous beam shown in Figure 7.13(a), which carries various vertical and horizontal loads. At the left-hand end A it is built-in to a foundation, at B and C it rests on rollers, and at D on a simple

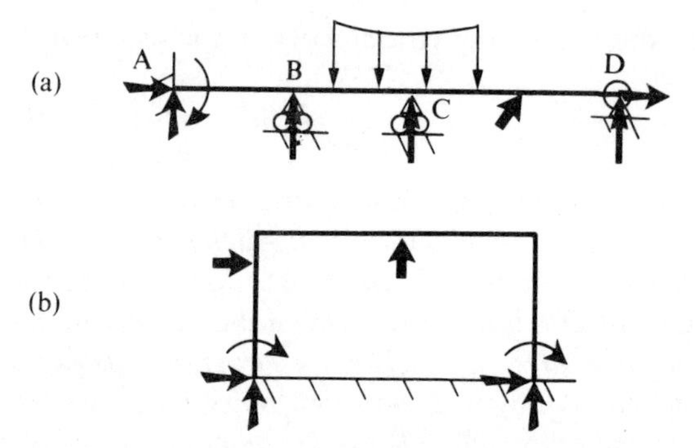

FIG. 7.13. (a) Loaded continuous beam. (b) Portal frame.

support. There are seven unknown reaction extered on the structure by its supports:

(i) a horizontal reaction at A,
(ii) a vertical reaction at A,
(iii) a moment at A,
(iv) a vertical reaction at B,
(v) a vertical reaction at C,
(vi) a horizontal reaction at D, and
(vii) a vertical reaction at D.

All the forces lie in the same plane, and so there are three independent equilibrium equations. Accordingly, the number of redundant forces is $7-3$, so that if four of the unknown reactions were known, the other three, and the internal stress resultants, could all be determined by statics.

Now consider a plane portal frame with fixed stanchion feet (Figure 7.13(b)). Again suppose that the loading is in the plane of the frame. There are then six reaction components, which determine the tension, shear force, and bending moment throughout the frame, and so the number of redundants is $6-3=3$. This may seem puzzling, because this frame was analysed in Example 7.3.3, and only two redundant forces were used. In that particular example, however, both the frame and the loading were symmetrical, and so symmetry gave us extra information, and told us that the reactions at the column feet were equal. Although the frame was three times redundant, this extra piece of information enabled us to reduce the number of unknown redundants to two.

If the same portal frame carries a more general loading, with load components perpendicular to its plane, the problem becomes three-dimensional. There are then six independent unknown reactions at each of the stanchion feet (three force components and three moment components at each), and so there are altogether 12 unknown forces and moments. There are six equilibrium equations, and therefore $12-6=6$ redundants.

A useful alternative approach is to think of how the structure must be modified to make it statically determinate. One can imagine it modified by introducing 'releases' which remove reactions and internal stress resultants, in the same way as earlier we thought of removing bars from frameworks. Each release removes one force resultant, and the number of releases that have to be put in to make a frame determinate is equal to the number of redundancies of its original form. Consider again the plane frame with fixed feet (Figure 7.14(a)), loaded in its own plane. If the built-in stanchion foot is

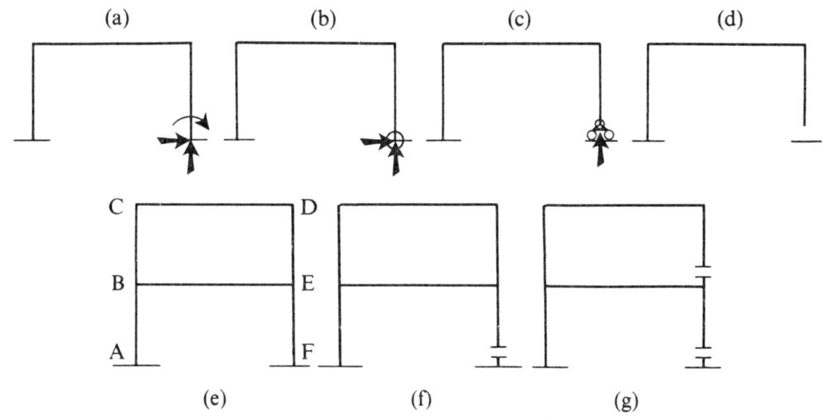

FIG. 7.14. (a) Portal frame with both feet fixed. (b) Portal frame (a), with one fixed foot replaced by a hinge. (c) Portal frame (a), with one fixed foot replaced by a hinged roller. (d) Portal frame (a), with one stanchion foot free. (e) Two-storey single-bay portal frame. (f) Portal frame (e), with cut in lower-storey right-hand stanchion. (g) Portal frame (e), with cuts in right-hand stanchion in both storeys.

replaced by a hinge (Figure 7.14(b)), one of the unknown reactions has disappeared, but the frame is still indeterminate. A second release removes the horizontal component of the reaction at the stanchion foot, and is equivalent to putting the foot on a roller (Figure 7.14(c)). Finally, a third release (Figure 7.14(d)) removes the vertical component. The frame is now statically determinate, and forces and moments within it can be found by statics. Three releases were required to bring it to this state, and so the original frame was three times redundant.

A complete cut through a plane frame releases three internal stress resultants, the tension, shear force, and bending moment. Consider the two-storey frame ABCDEF shown in Figure 7.14(e). One cut in EF (Figure 7.14(f)) releases three stress resultants, and makes the forces within AB and EF determinate, but leaves those in BCDE indeterminate. A second cut, in the 'ring' BCDE, makes the structure determinate, but releases three more stress resultants. Accordingly, the original frame was $2 \times 3 = 6$ times redundant.

7.6. Redistribution of forces in nonlinear structures

Most of the structures we have so far considered have had linear relationships between stress resultants, such as tension and bending moment, and their corresponding deformation quantities, such as elongation and curvature. Reactions and internal stress resultants are then proportional to loads. If the relationships are nonlinear, this is no longer so.

Example 7.6.1. A uniform beam is built-in at one end and rests on a simple support at the other end (Figure 7.15(a)). Bending moment and curvature are related in the way shown in Figure 7.15(b), which is an idealization of the observed behaviour of steel and lightly reinforced concrete beams. The beam carries a uniformly distributed load w per unit length, whose intensity is steadily increased, starting at zero. What is the distribution of bending moment in the beam, and how does the deflection at the centre of the beam depend on w?

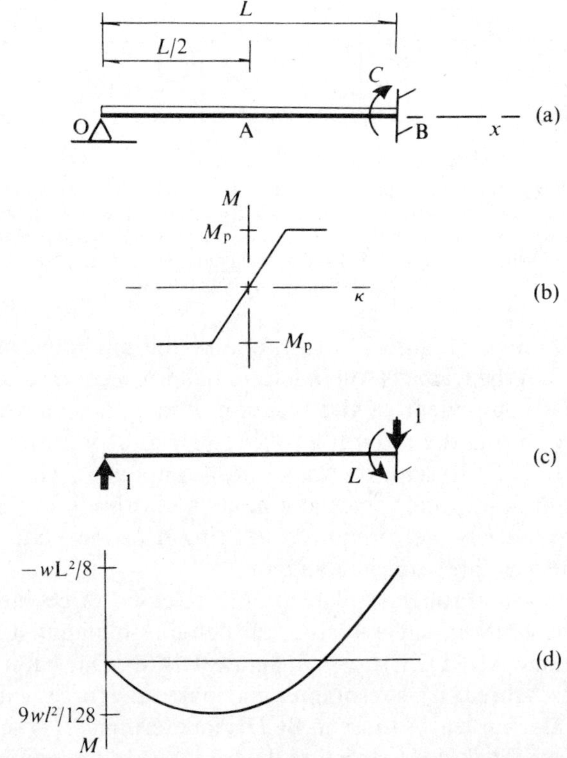

FIG. 7.15. (a) Uniformly loaded beam, built-in at one end and simply supported at the other. (b) Moment–curvature relationship. (c) Unit end reaction at left-hand support. (d) Distribution of bending moment.

If the end simple support were removed, the beam would be statically determinate. As it is, it is once redundant. Choose as the redundant the end fixing moment C, so that the moments and forces in the beam can be considered as the sum of those that would exist if C were zero, and those that would exist if C were zero but w were not. Then

$$M=\tfrac{1}{2}wx(L-x)-Cx/L, \tag{7.6.1}$$

where x is the distance from the left-hand end.

When w is small, the moment will be small everywhere, so that $|M|<M_{\text{p}}$, and the moment and the curvature are related by

$$M=F\kappa,$$

where F is a constant flexural rigidity, and so

$$\kappa=\{\tfrac{1}{2}wx(L-x)-Cx/L\}/F. \tag{7.6.2}$$

Continuing the analysis in the usual way, a system of moments, in equilibrium with zero external forces and moments except at the supports, corresponds to Figure 7.15(c), and has

$$M^{(1)}=x \tag{7.6.3}$$

Choosing this as the force and moment system in the virtual work equation, and choosing the actual curvatures κ and displacements,

$$0=\int_0^L M^{(1)}\kappa\,\mathrm{d}x, \tag{7.6.4}$$

$$0=\tfrac{1}{24}wL^4-\tfrac{1}{3}CL^2 \tag{7.6.5}$$

and so

$$\begin{aligned} C&=wL^2/8\\ M&=\tfrac{3}{8}wLx-\tfrac{1}{2}wx^2. \end{aligned} \tag{7.6.6}$$

The distribution of moment is shown in Figure 7.15(d). Its maximum value is at $x=3L/8$, where M is $9wL^2/128$, and its minimum at $x=L$, where $M=-C=-wL^2/8$.

The deflection at the centre is found by the usual unit load method. The distribution of moment

$$M^{(2)}=[x-\tfrac{1}{2}L], \tag{7.6.7}$$

where [ ] is as usual a Macauley bracket, is in equilibrium with unit vertical load at the centre. Choosing this as the equilibrium set in the virtual work equation, and the actual curvature and displacements as the compatible set,

$$v_{\text{a}}=\int_0^L M^{(2)}\kappa\,\mathrm{d}x=-wL^4/192. \tag{7.6.8}$$

If w is increased, the bending moment somewhere in the beam will eventually reach M_p or $-M_p$, at which curvature can go on increasing without any further change in moment. Equation (7.6.2) then no longer applies, at least not over the whole beam, and further analysis is needed. Since $|9wL^2/128|$ is less than $|-wL^2/8|$, this condition is reached first at the end $x = L$, when

$$-wL^2/8 = -M_p$$
$$w = 8M_p/L^2. \tag{7.6.9}$$

When w is increased beyond this value, the moment at B can no longer increase in absolute value, but remains fixed at $-M_p$. What then happens is that a large curvature develops at B, so that the deflected shape of the beam is that shown in Figure 7.16(a). A *plastic hinge* is formed, in which a large

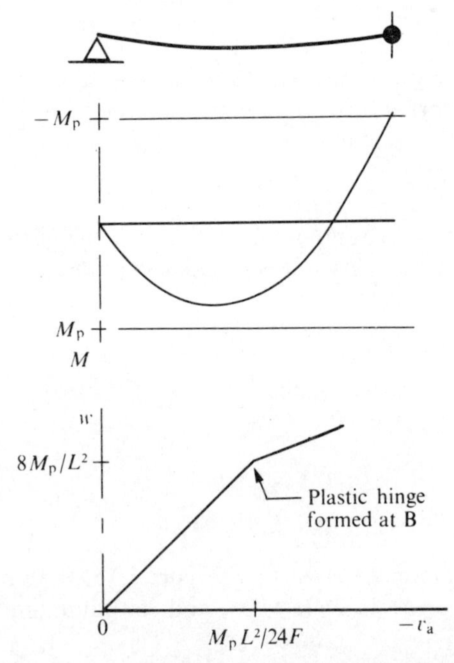

FIG. 7.16. (a) Deflection of beam of Figure 7.15, after a plastic hinge has formed at the right-hand support. (b) Corresponding distribution of bending moment. (c) Relation between central deflection and load intensity.

curvature occurs over a short length. Hinges of this kind are observed in tests on steel structures: you can observe them in a desk-top experiment by making the beam of Figure 7.15 from a length of wire, resting one end on a pencil and clamping the other between pliers or coins, and applying load with your fingers. Plastic hinges are indicated in diagrams in the way shown in Figure 7.16(a). A plastic hinge transmits moment: the solid circle distinguishes it from a free hinge, represented by an open circle.

Now that the moment at B is fixed at $-M_p$

$$C = M_p \tag{7.6.10}$$

and

$$M = \tfrac{1}{2}wx(L-x) - M_p x/L, \tag{7.6.11}$$

which has a maximum

$$wL^2/8 - \tfrac{1}{2}M_p + \tfrac{1}{2}M_p^2/wL^2$$

at $x = \frac{1}{2}L - M_p/wL$. This distribution is plotted in Figure 7.16(b).

Equation (7.6.4) still applies, but we must now remember that only part of the beam is elastic. In the elastic segment κ is still given by eqn (7.6.2), with C replaced by M_p. The plastic hinge is at $x = L$, where $M^{(1)} = L$. Accordingly

$$0 = \int_{\substack{\text{elastic}\\\text{segment}}} M^{(1)}\kappa\,\mathrm{d}x + \int_{\substack{\text{plastic}\\\text{hinge}}} M^{(1)}\kappa\,\mathrm{d}x \tag{7.6.12}$$

$$= \int_0^L \frac{x}{F}\left\{\frac{1}{2}wx(L-x) - M_p x/L\right\}\mathrm{d}x + L\int_{\substack{\text{plastic}\\\text{hinge}}} \kappa\,\mathrm{d}x \tag{7.6.13}$$

and so

$$L\int_{\substack{\text{plastic}\\\text{hinge}}} \kappa\,\mathrm{d}x = -\frac{1}{F}(wL^3/24 - M_p/3L). \tag{7.6.14}$$

The integral of κ dx over the hinge is an angle, the rotation of the hinge, the relative rotation between the still-linear portion of the beam to the left and the fixed end to the right. The deflection at A can be found in the same way as before

$$v_a = \int_{\text{beam}} M^{(2)}\kappa\,\mathrm{d}x \tag{7.6.15}$$

$$= \int_{\substack{\text{elastic}\\\text{segment}}} M^{(2)}\kappa\,\mathrm{d}x + \int_{\substack{\text{plastic}\\\text{hinge}}} M^{(2)}\kappa\,\mathrm{d}x \tag{7.6.16}$$

$$= \int_{\frac{1}{2}L}^{L} \left(x - \frac{1}{2}L\right)\frac{1}{F}\left\{\frac{1}{2}wx(L-x) - M_p x/L\right\}\mathrm{d}x + \frac{1}{2}L\int_{\substack{\text{plastic}\\\text{hinge}}} \kappa\,\mathrm{d}x \tag{7.6.17}$$

$$= -\frac{5}{384}wL^4/F + \frac{1}{16}M_p L^2/F, \tag{7.6.18}$$

using eqn (7.6.14).

The relationship between load and deflection at A is plotted in Figure 7.16(c). As we would expect, the deflection increases much more rapidly once a hinge has formed at B.

The distribution of bending moment after a plastic hinge has formed at B is shown in Figure 7.16(b). To start with the remainder of the beam is still elastic. Eventually, however, the maximum sagging moment also reaches the yield value. This occurs at a larger load, when

$$M_p = wL^2/8 - \tfrac{1}{2}M_p + \tfrac{1}{2}M_p^2/wL^2, \tag{7.6.19}$$

that is, when

$$w = (6+4\sqrt{2})M_p/L^2 = 11{\cdot}66\ M_p/L^2. \tag{7.6.20}$$

There are then two plastic hinges (Figure 7.17(a)), the first one still at B and a second one at the point at which the maximum sagging moment first reaches M_p, at

$$x = (\sqrt{2}-1)L = 0{\cdot}414L. \tag{7.6.21}$$

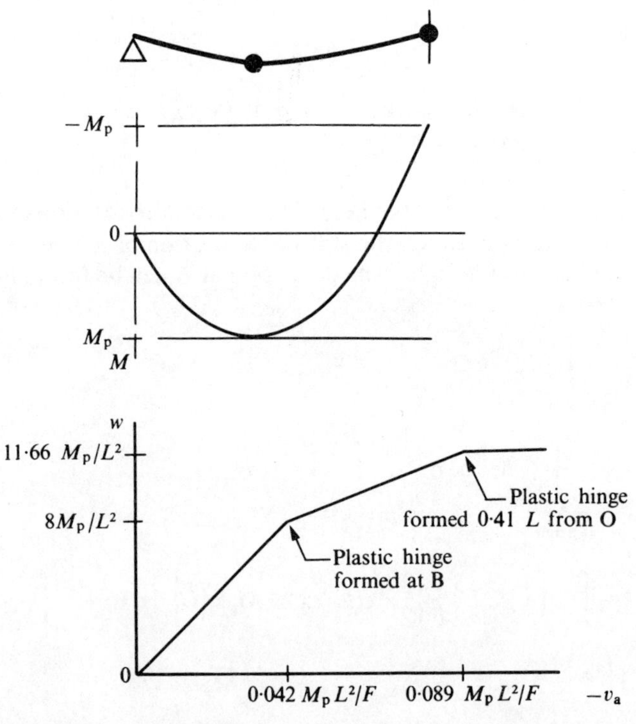

FIG. 7.17. (a) Deflection of beam of Figure 7.15, after a second plastic hinge has formed. (b) Corresponding distribution of bending moment. (c) Complete relation between central deflection and load intensity.

A collapse mechanism has been formed, and the beam can continue to deform without any further increase of load, the deflection developing by increasing rotation at the two hinges. Figure 7.17(b) shows the bending moment distribution.

This kind of behaviour is characteristic of structures which have moment–curvature relations that display plastic yield, like the one in Figure 7.15(b). The most important thing an engineer needs to know about such a structure is the value of the collapse load at which deflections will become very large. The analysis just completed found that load by following through the loading process, observing the formation of plastic hinges, and modifying the analysis accordingly. It turns out that the collapse load can be found by a much simpler analysis, and this is explained in Chapter 9.

You will notice in this problem that the distribution of bending moment alters as the loading continues. The maximum sagging moment is not always at the same point, and the ratio between the maximum hogging moment and the maximum sagging moment falls from 16/9 to 1. Redistribution of this kind occurs in many nonlinear structures.

7.7. Problems

1. Using the virtual work method, find
   (i) the end deflection of a uniform cantilever (length L, flexural rigidity F) produced by a concentrated end load P.
   (ii) the central deflection of a simply supported uniform beam (length L, flexural rigidity F), produced by a uniformly distributed load wL.

2. Use virtual work to solve Problem 3 in Section 5.6.

3. A curved bar has a uniform flexural rigidity F and the shape of a semi-circle of radius R. If equal and opposite loads P are applied to the ends, what are the consequent relative deflections and rotations of the ends? If equal and opposite moments C are applied, what are the consequent relative deflections and rotations of the ends?

4. A uniform continuous beam of length $2L$ and flexural rigidity F rests on three fixed supports A, B, and C, one at each end and one at the centre. One span, AB, carries a uniformly distributed load wL; the other span, BC, is unloaded.
   (i) Find the curvature of the beam at a point distant x from A, in terms of w and a redundant reaction R at C.
   (ii) Find a self-equilibrating bending moment system in equilibrium with zero external loads except at the supports.
   (iii) Thence use virtual work to find R and the other two support reactions.
   (iv) Find the deflection of the centre of the loaded span.

5. Figure 7.18 shows a square portal frame of height and span L. All its members have the same uniform cross-section, and the stanchion feet are hinged to a rigid foundation. A uniformly distributed load of intensity p per unit length is applied to the left-hand stanchion over its whole height. Find the position and magnitude of the greatest bending moment set up in the frame by the load. Assume that the bending

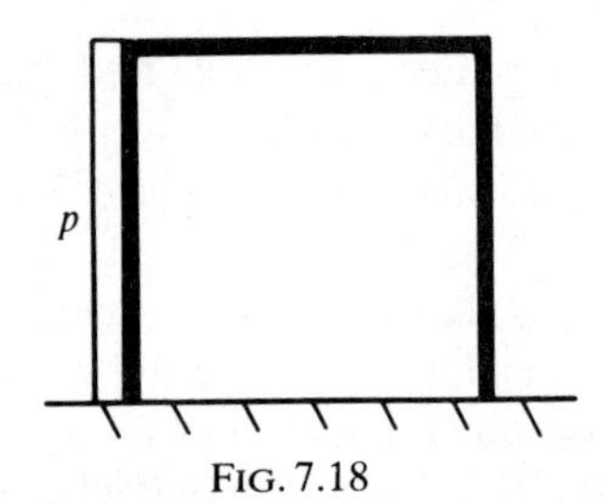

FIG. 7.18

moment and curvature are linearly related, and that only bending deformations need to be taken into account.

6. A symmetrical two-hinged arch bridge (Figure 7.19) is parabolic in profile. The span between the abutment hinges is L, and the rise of the centre above a line between the hinges is h. The flexural stiffness at a point x to the right of the centreline is $F(x)$.

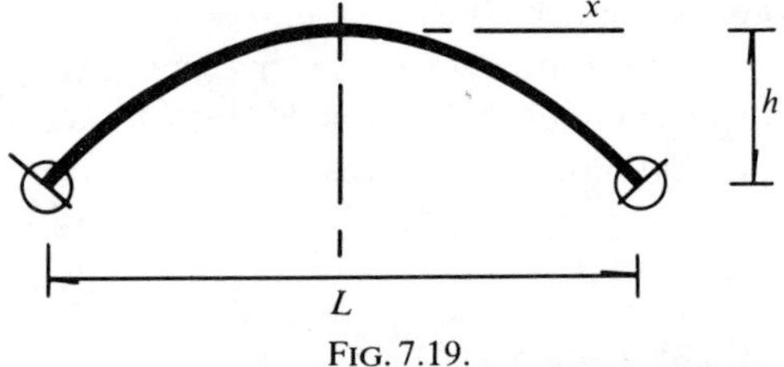

FIG. 7.19.

The arch carries a central vertical load P. Derive an expression from which the induced horizontal abutment thrust Q could be calculated, given L, h, and the function $F(x)$. Then simplify the resulting expression by assuming that the flexural rigidity F is uniform and that the inclination of the arch rib to the horizontal is everywhere small, and thence show that in that case the abutment thrust is $25PL/128h$. Sketch the resulting bending moment distribution.

7. The pin-jointed plane framework shown in Figure 7.20 has twelve identical bars and is stress-free when unloaded. The bars have a uniform cross-sectional area A, and are made from a material which obeys Hooke's law and has Young's modulus E and linear thermal expansion coefficient α. Find the tension in each bar of the framework when the temperature in bar GC is raised by θ. Calculate this tension when A is $1000\ \text{mm}^2$, θ is 50 K, and the framework is made of steel ($E = 210\ \text{kN/mm}^2$, $\alpha = 1{\cdot}1 \times 10^{-5}\ \text{K}^{-1}$).

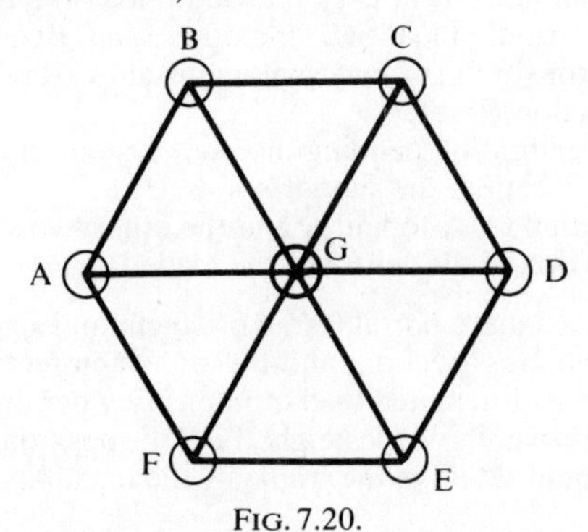

FIG. 7.20.

8. Figure 7.21 shows a pin-jointed plane framework. The bar properties are listed below. At 10°C the bars are of the correct length except for BC, which is 1 mm too long. The framework is then assembled, the temperature rises to 20°C, and equal and opposite outward forces of 10 kN are applied at B and D. What is then the tension in BD?

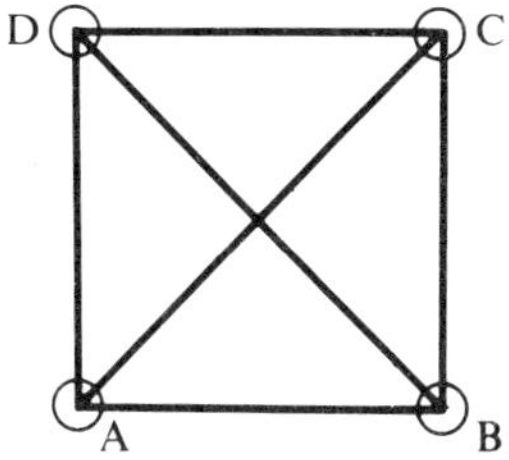

FIG. 7.21.

Bar	Material	Young's modulus E (kN/mm^2)	Linear thermal expansion coefficient (K^{-1})	Cross-section (mm^2)
AB, BC, CD, DA	brass	105	$2{\cdot}0\times10^{-5}$	100
AC	aluminium	70	$2{\cdot}2\times10^{-5}$	500
BD	steel	210	$1{\cdot}1\times10^{-5}$	50

9. A square portal frame of height and span L has its stanchion feet built-in to a rigid foundation. Its flexural rigidity is uniformly F, and its elements have a linear thermal expansion coefficient α. Find the horizontal reactions and moments induced at the stanchion feet by a temperature rise of θ.

10. Determine the number of redundant forces (or moments) for each of the plane frames in Figure 7.22, which have rigid joints except where a circle indicates a hinged joint.

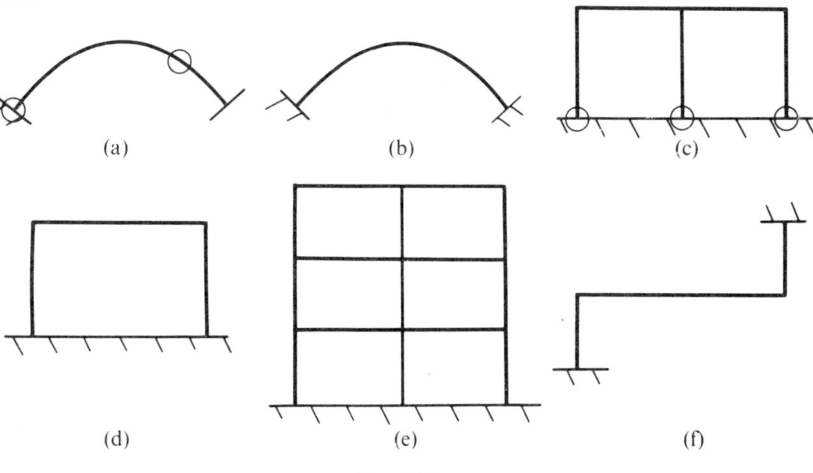

FIG. 7.22.

8. Linear structures

8.1. Introduction

IF a structure is loaded, it responds to the load: internal forces appear within it, and its elements bend, stretch, and twist, so that it deflects. Many structures show a linear relationship between the loading and the structural response, so that (for example) the deflection is proportional to the load. This chapter is concerned with some special results that apply only to linear structures. Linear response is important in engineering practice because most of the materials that structures are made from behave almost linearly at working loads. Thus, a column made from steel or aluminium will shorten under load by an amount almost exactly proportional to the load, as long as the yield stress of the metal is not reached. If the column is made of concrete, or brickwork, or glass-reinforced plastic, the response will not quite be linear, even at low loads, but an idealization which represents its response as linear will nevertheless give us a good approximation to its behaviour. Structures made from these materials will behave almost linearly at loads up to their working loads, and the results of this chapter will then apply to them.

8.2. Superposition

The equilibrium equations are linear. If a set of forces is in equilibrium, and we multiply all the forces by 2 or 100 or -1 (which is equivalent to reversing their signs), or indeed by any multiplier λ, then they will still be in equilibrium. The fact that the equations are linear also tells us that if one set of forces and moments acting on a body together keep it in equilibrium, and a second independent set of forces and moments would also keep it in equilibrium, then if the two sets are added together, so that they both act at once, the body will still be in equilibrium. For instance, the rectangular body in Figure 8.1 is in equilibrium either under the set of forces (A) shown in Figure 8.1(a), or under the set of forces (B) in Figure 8.1(b). If the sets (A) and (B) are superposed, as in Figure 5.1(c), the resulting set of forces will automatically satisfy the equilibrium conditions. Similarly, if the forces of set (B) are multiplied by 4, and then added to set (A), as in Figure 8.1(d), the equilibrium conditions will again automatically be obeyed.

In this limited sense, systems of forces automatically obey a *principle of superposition*: if system (A) is in equilibrium, and system (B) is in equilibrium, (A) and (B) superposed will be in equilibrium. The principle will apply to any statically determinate structure, and in that case to internal forces as well as external loads. If one set of loads (A) applied to a statically determinate framework induces a tension of +10 in one of its bars, and a

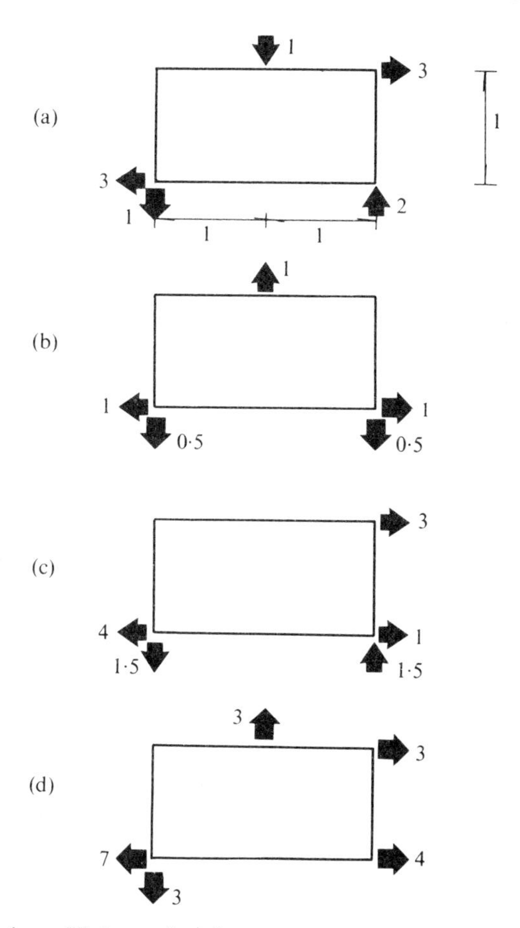

FIG. 8.1. (a) Body in equilibrium under a first set of forces (A). (b) Body in equilibrium under a second set of forces (B). (c) Body in equilibrium under (A) and (B) acting together. (d) Body in equilibrium under (A) and four times (B) acting together.

second set (B) induces a tension of -7 in the same bar, then (A) and (B) together will induce a tension of $10+(-7)=3$; the difference between (A) and (B), symbolically (A $-$ B), will induce a tension of $10-(-7)=17$; three times (A) superposed on (B) will induce a tension of $(3\times 10)+(-7)=23$, and so on. For statically determinate structures this is true generally, except when the deformations induced by the loads are so great that they alter the whole geometry of the structure.

Exactly as the equations describing statics are linear, so are the equations describing the kinematics of small deformations, the equations that were derived in Chapter 5. In the analysis of statically indeterminate structures we use in addition generalized stress–strain relations, such as those that relate bar elongation to bar tension, and beam curvature to bending moment. Relations of this kind are not always linear. If, but only if, all the relevant relations of this kind that apply to a particular structure are indeed linear, then a more general principle of superposition holds:

If one set of loads $\mathbf{L}_A$ *induces in a structure a certain response* $\mathbf{R}_A$*, expressed by internal forces, moments, deformations, and so on at different points in the structure, and a second set of loads* $\mathbf{L}_B$ *induces another response* $\mathbf{R}_B$*, then if the two sets act together,* $\mathbf{L}_A + \mathbf{L}_B$*, the response is* $\mathbf{R}_A + \mathbf{R}_B$*, the sum of the responses to the two sets acting separately. Generalizing this, if m times* $\mathbf{L}_A$ *acts with n times* $\mathbf{L}_B$*, so that the loading is* $m\mathbf{L}_A + n\mathbf{L}_B$*, the response is* $m\mathbf{R}_A + n\mathbf{R}_B$*, m times the response to* $\mathbf{L}_A$ *plus n times the response to* $\mathbf{L}_B$.

To illustrate this, consider a uniform beam, built-in at both ends, with a linear relationship between its bending moment and its curvature, so that it obeys the conditions for the wider principle of superposition to hold. Its flexural rigidity F is constant along its length, and its length is L. In Figure 8.2(a) it carries a central concentrated load W. The bending moment

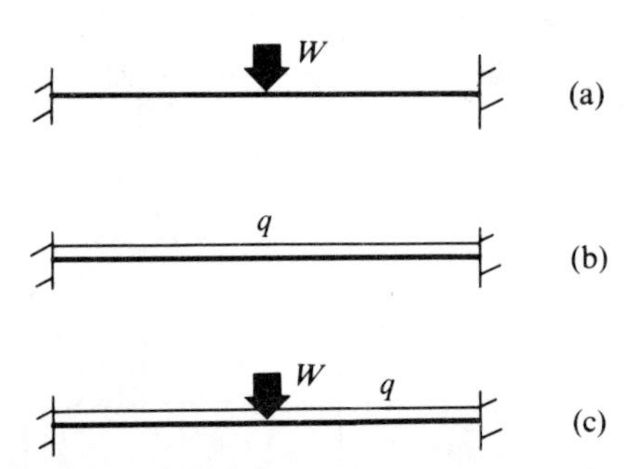

FIG. 8.2. (a) Uniform beam under central concentrated load. (b) Uniform beam under distributed load. (c) Uniform beam under central concentrated load and distributed load acting together.

induced at the left-hand end can be shown to be $-WL/8$, and the central deflection is $WL^3/192F$. In Figure 8.2(b) the same beam carries a distributed load of intensity q per unit length, and the end moment is then $-qL^2/12$, and the central deflection $qL^4/384F$. Having these results, we can say without any further analysis that if the two loadings act together (Figure 8.2(c)) the end moment is $-(WL/8 + qL^2/12)$ and the central deflection $WL^3/192F + qL^4/384F$, whatever the values of W and q. It would not be possible to make a similar statement if the moment–curvature relation were nonlinear.

Superposition is useful because it makes it possible to analyse the effects of complex loadings by superposing the effects of simple loadings, which are easier to analyse individually. For very simple linear structures, like the one shown in Figure 8.2, the responses to simple loadings are so often needed by structural engineers that they are available in tabulated form in handbooks, and superposition can be used to construct responses to complex loadings (see Problem 8.4). The next section shows that it can also be used to simplify calculation, for structures which possess some kinds of symmetry.

8.3. Superposition and symmetry

When one studies symmetric structures, superposition and symmetry can sometimes be used together to reduce the amount of calculation required. There is no systematic method, but the ideas involved are illustrated by the following example, and additional problems will be found at the end of the chapter.

Figure 8.3(a) shows a linear symmetrical portal frame, with built-in stanchion feet. Initially the loads and reactions on the frame are all zero, and

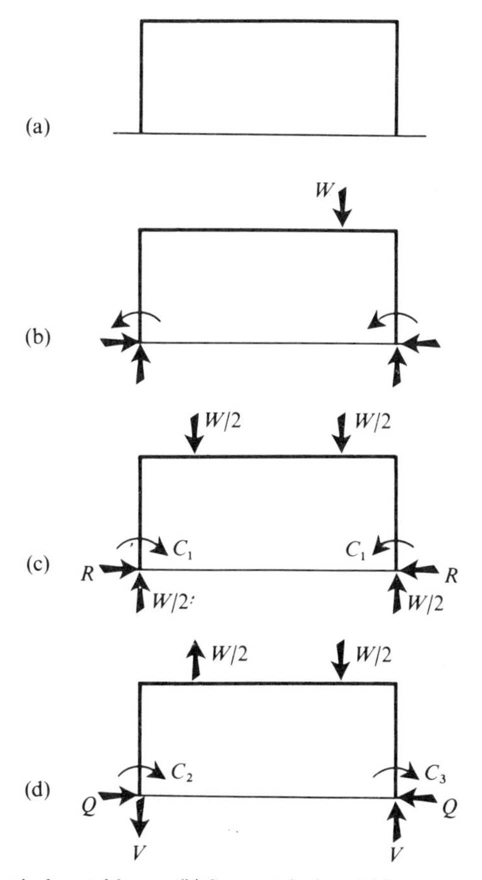

FIG. 8.3. (a) Symmetrical portal frame. (b) Symmetrical portal frame, carrying a concentrated load. (c) Symmetrical portal frame, under symmetrical loading (C). (d) Symmetrical portal frame, under anti-symmetric loading (D).

so it is free of bending moment. In Figure 8.3(b) it carries a concentrated load W asymmetrically placed on the beam. The frame is three times redundant, and because the load is asymmetric none of the redundant forces can be found from symmetry.

Consider the same frame loaded in the ways shown in Figures 8.3(c) and 8.3(d). If we superpose these loadings, we get back the asymmetric loading of Figure 8.3(b). Reversing the superposition process, we can think of the first loading as divided into the loading (C) of Figure 8.3(c) and the loading (D) of Figure 8.3(d). The frame is of course still three times redundant, but loading (C) is symmetric about a vertical axis through the midpoint of the beam, and the frame is symmetric in the same way, and so the vertical reactions at the column feet must be equal. Their sum is W, by vertical equilibrium, and so each of them is $\frac{1}{2}W$. The redundant horizontal reaction R and fixing moment C_1 are not determined by statics, and so a solution for loading (C) still requires a solution for two independent redundant forces, by the method of Chapter 7.

Consider next the loading (D), which is anti-symmetric. At the left-hand support the horizontal reaction is Q, say, to the right, Figure 8.3(d), the vertical reaction is V downwards, and the fixing moment is C_2, clockwise. At the right-hand support the horizontal reaction must be Q to the left (otherwise the frame cannot be in equilibrium), and the vertical reaction must be V upwards; the fixing moment is C_3, say. The bending moment at the centre of the beam is M^*. Now draw Figure 8.3(d) on a sheet of tracing paper, and turn the paper over, left-to-right. What you will see from the other side of the paper is Figure 8.4(a): the frame is the same but the loads have been reversed, so that the upward load is on the right and the

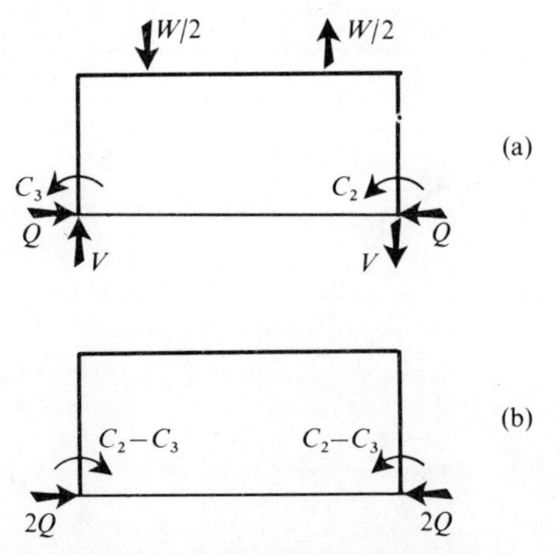

FIG. 8.4. (a) Loading (D) in Figure 8.3(d), seen from the other side of the frame; loading (E). (b) Superposition of loadings (D) and (E).

downward load on the left, and the reactions and moments are as shown. If loading (D) produces the reactions and bending moments of Figure 8.3(d), the reversed loading (E) in Figure 8.4(a) produces the reactions and bending moments shown there.

Now imagine loadings (D) and (E) superposed. The result is illustrated in Figure 8.4(b). The external loads have cancelled each other, and so have the vertical reactions V, but the resultant horizontal reaction is $2Q$ at each stanchion foot, and the fixing moments are each of magnitude $C_2 - C_3$. The bending moments at the centre of the beam have added, so that the bending moment is now $2M^*$. The frame, though, has no load on it, and we took as our starting point an unloaded frame which has no reactions or moments. It follows that the reactions and moments in Figure 8.4(b) must in fact be zero, that

$$2Q = 0,$$

$$C_2 - C_3 = 0,$$

$$2M^* = 0,$$

and therefore

$$Q = M^* = 0$$

and

$$C_2 = C_3.$$

Originally we described the reactions on the frame under anti-symmetric loading in terms of four quantities, Q, V, C_2, and C_3, and could determine the bending moments within the frame in terms of them. Symmetry has shown us that Q is zero, that C_2 and C_3 are equal, and that there is zero bending moment at the centre of the beam (which gives us another relation between W, V, and C_2, by taking moments). Only one of the four quantities is independent. Given that one, the others can be found by statics, and it follows that in an analysis of the anti-symmetric loading (D) we need consider only one independent redundant quantity, instead of three.

In this way we have used symmetry to reduce the analysis of a three times redundant structure under asymmetric loading to two separate problems, one for symmetric loading (C), with two independent redundants, and the other for anti-symmetric loading (D), with one independent redundant. On the way we have derived useful auxiliary information about the anti-symmetric case. The amount of effort needed to solve the original problem by this route is less than that by a direct attack by the techniques of Chapter 7, because solution of a structure with n independent redundants leads to a set of n equations in n unknowns, and the number of calculations needed to solve such a set is proportional to n^3, and so the calculation effort has been reduced in the ratio $3^3 : 2^3 + 1^3$, by a factor of three.

8.4. The reciprocal theorem

A useful theorem about linear structures was first proved by Maxwell and generalized by Betti. The statement of the theorem is this:

Suppose a number of forces $P_1^{(1)}, P_2^{(1)}, \ldots, P_n^{(1)}$ to act simultaneously on a linear structure, and that the corresponding displacements along the lines of action of these forces are respectively $u_1^{(1)}, u_2^{(1)}, \ldots, u_n^{(1)}$. If these forces are replaced by a second system $P_1^{(2)}, P_2^{(2)}, \ldots, P_n^{(2)}$, acting at the same points and in the same directions as the first system, the corresponding displacements being $u_1^{(2)}, u_2^{(2)}, \ldots, u_n^{(2)}$, then

$$P_1^{(1)}u_1^{(2)} + P_2^{(1)}u_2^{(2)} + \cdots + P_n^{(1)}u_n^{(2)} = P_1^{(2)}u_1^{(1)} + P_2^{(2)}u_2^{(1)} + \cdots + P_n^{(2)}u_n^{(1)}. \tag{8.4.1}$$

Imagine, for example, a beam ABCD, built-in at A and simply supported at D (Figure 8.5(a)). Suppose the first loading to be a unit vertical load at B

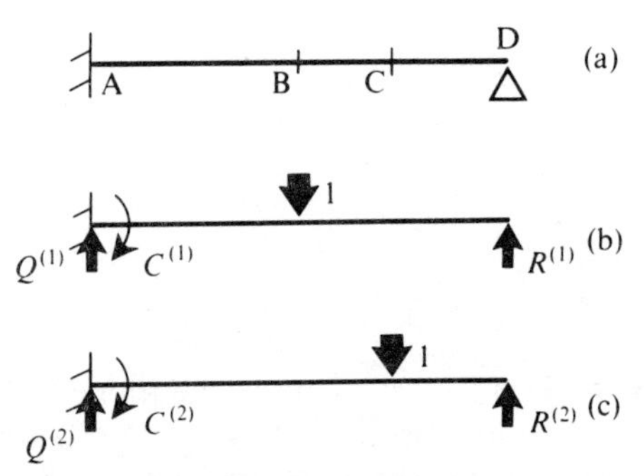

FIG. 8.5. (a) Beam ABCD. (b) Beam ABCD, carrying unit vertical load at B. (c) Beam ABCD, carrying unit vertical load at C.

(Figure 8.5(b)). Deflections $v_B^{(1)}$ and $v_C^{(1)}$ occur at B and C, and reactions are induced at the supports, a reaction $Q^{(1)}$ and a moment $C^{(1)}$ at A and a reaction $R^{(1)}$ at D. All the forces and moments acting on the beam, and their corresponding displacements, are listed in Table 8.1.

TABLE 8.1

Point	Force (moment)	Corresponding displacement (rotation)
A	$Q^{(1)}$	0
A	$C^{(1)}$	0
B	1	$v_B^{(1)}$
C	0	$v_C^{(1)}$
D	$R^{(1)}$	0

Suppose the second loading to be a unit vertical force at C (Figure 8.5(c)). Once again, deflections occur at B and C, but they have different values $v_B^{(2)}$ and $v_C^{(2)}$, and reactions are induced at the supports, $Q^{(2)}$ and $C^{(2)}$ at A and $R^{(2)}$ at D. The forces and moments acting on the beam in this loading are listed in Table 8.2.

TABLE 8.2

Point	Force (moment)	Corresponding displacement (rotation)
A	$Q^{(2)}$	0
A	$C^{(2)}$	0
B	0	$v_B^{(2)}$
C	1	$v_C^{(2)}$
D	$R^{(2)}$	0

Now apply the theorem, taking care to include all the loads. Equation (8.4.1) gives

$$Q^{(1)} . 0 + C^{(1)} . 0 + 1 . v_B^{(2)} + 0 . v_C^{(2)} + R^{(1)} . 0$$
$$= Q^{(2)} . 0 + C^{(2)} . 0 + 0 . v_B^{(1)} + 1 . v_c^{(1)} + R^{(2)} . 0$$

and so

$$v_B^{(2)} = v_C^{(1)}. \tag{8.4.2}$$

In words, the vertical deflection at B produced by a unit vertical load at C is the same as the vertical deflection at C produced by a unit vertical load at B. This is by no means obvious, and in fact rather surprising, but similar results turn out to apply to any linear structure. They apply whether or not the structure is statically determinate, and whether or not its elements are uniform or symmetric, and only require that it be made from a linear material.

The proof of the theorem is quite straightforward. We shall only prove it for a pin-jointed framework, but a similar proof applies to any linear structure. In the proof we need first of all the statement of virtual work, eqn (6.2.23), that

$$\sum_{\text{all joints}} (\text{load}) \begin{pmatrix} \text{corresponding} \\ \text{displacement} \end{pmatrix} = \sum_{\text{all bars}} T_{ij} e_{ij}, \tag{8.4.3}$$

where the bar tensions T_{ij} are in equilibrium with the loads, and the elongations are compatible with the displacements. Secondly, since each bar now has a linear relationship between its tension and its elongation, we can relate the elongation of a bar IJ to its tension by

$$e_{ij} = k_{ij} T_{ij}, \tag{8.4.4}$$

where k_{ij} is a constant for each bar, called the compliance.

In the first loading, the loads are $P_1^{(1)}, \ldots, P_n^{(1)}$, and they induce bar tensions $T_{ij}^{(1)}$; the corresponding joint displacements are $u_1^{(1)}, \ldots, u_n^{(1)}$, and the elongations $e_{ij}^{(1)}$. In the second loading, the loads are $P_1^{(2)}, \ldots, P_n^{(2)}$, the bar tensions $T_{ij}^{(2)}$, the joint displacements $u_1^{(2)}, \ldots, u_n^{(2)}$, and the elongations $e_{ij}^{(2)}$. Apply the virtual work statement, eqn (8.4.3), choosing the loads and bar forces in equilibrium from loading (1) and the compatible displacements and elongations from loading (2); then

$$P_1^{(1)} u_1^{(2)} + \cdots + P_n^{(1)} u_n^{(2)} = \sum_{\text{all bars}} T_{ij}^{(1)} e_{ij}^{(2)}. \tag{8.4.5}$$

Apply virtual work a second time, choosing the loads and bar forces in equilibrium from loading (2) and the compatible displacements and elongations from loading (1); then

$$P_1^{(2)} u_1^{(1)} + \cdots + P_n^{(2)} u_n^{(1)} = \sum_{\text{all bars}} T_{ij}^{(2)} e_{ij}^{(1)}. \tag{8.4.6}$$

So far equations (8.4.5) and (8.4.6) apply whether or not a bar elongation is linearly related to its tension. Now substitute from eqn (8.4.4) into the right-hand side of eqn (8.4.5); it becomes

$$\sum_{\text{all bars}} T_{ij}^{(1)} k_{ij} T_{ij}^{(2)},$$

the sum over all bars of the compliance multiplied by the tension in the first loading multiplied by the tension in the second loading. Similarly, the right-hand side of eqn (8.4.6) is

$$\sum_{\text{all bars}} T_{ij}^{(2)} k_{ij} T_{ij}^{(1)},$$

which is the same. Since the right-hand sides of eqns (8.4.5) and (8.4.6) are equal, the left-hand sides must be equal too, and the theorem is proved.

Be careful not to confuse the reciprocal theorem with virtual work. Virtual work relates one set of compatible quantities, which describe deformation, to another set of quantities which describe forces in equilibrium; the deformations do not need to be the ones produced by the forces, which can indeed be wholly imaginary. The reciprocal theorem, on the other hand, relates one loading, and the actual deformations it produces, to a second loading, and the actual deformation it produces. Unlike virtual work, it applies only to linear structures.

A particularly useful application of the theorem is to the indirect use of structural models. This makes it possible to deduce forces within a loaded structure from experiments in which only deformations are measured, and not forces. The general principle—Müller–Breslau's principle—is to apply to the unloaded structure a displacement corresponding to the unknown force, and then to measure the consequent displacements of the points at which the loads on the structure act. An application of the theorem then gives the unknown force in terms of the loads, the applied displacement, and the measured displacement. It is not obvious that this can be done, but the following simple example shows how it comes about.

Consider a two-span continuous beam resting on three fixed simple supports, and the problem of determining a graph which describes how the left-hand support reaction varies as a unit load moves along the beam. Such a graph is called an influence line. Suppose two load systems to load the structure in turn. In the first loading, a unit load acts at a general point X on the beam (Figure 8.6(a)); this is the loading for which we are seeking the end

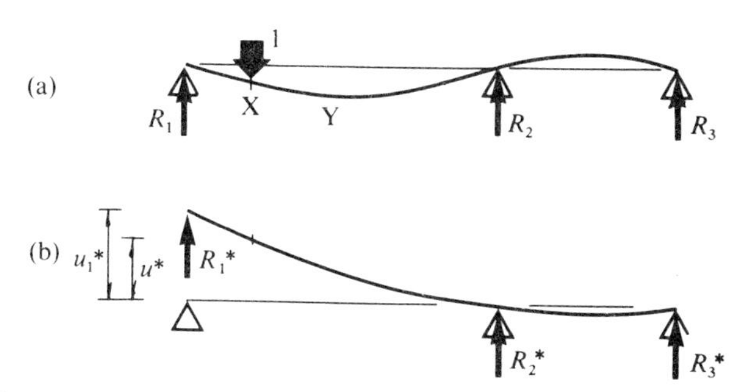

FIG. 8.6. (a) Two-span continuous beam, carrying unit load at X. (b) Two-span continuous beam, with left-hand end displaced vertically upwards.

reaction R_1 as a function of the position X. In the second loading, the beam has been displaced vertically at the left-hand support, a displacement corresponding to the unknown reaction. A force is required to do this, of course, and reactions are induced at the other two supports, but no other loads act on the beam (Figure 8.6(b)).

Table 8.3 lists the points at which loads act on the beam, and the corresponding loads and displacements in the two loading systems.

Applying the reciprocal theorem,

$$R_1 \,.\, u_1^* + (-1)u^* + R_2 \,.\, 0 + R_3 \,.\, 0 = R_1^* \,.\, 0 + 0 \,.\, u + R_2^* \,.\, 0 + R_3^* \,.\, 0,$$

that is

$$R_1 = \frac{u^*}{u_1^*}, \tag{8.4.7}$$

TABLE 8.3

Point	System (1) Force	System (1) Corresponding displacement	System (2) Force	System (2) Corresponding displacement
1	R_1	0	R_1^*	u_1^*
X	-1	$-u$	0	u^*
2	R_2	0	R_2^*	0
3	R_3	0	R_3^*	0

so that the reaction at support 1 due to unit load at X is equal to the ratio of the deflection of X to the deflection of 1 in loading (2), where a displacement corresponding to the unknown reaction has been introduced into the unloaded structure. The deflection under loading (2) is therefore an influence line for the reaction at 1 under loading (1). This is somewhat surprising, because it looks at first sight as if what happens under loading (2) is so different from what happens under loading (1) that one can tell us nothing about the other, but the result is correct.

If system (2) is modelled to some convenient scale, then as long as geometric similarity is maintained the ratio of u^* to u_1^* will be the same for the model as for the system (2) deformation of the prototype. This means that we can determine the support reaction for a full-scale beam by making a model to some convenient scale, deflecting the left-hand support through a measured distance in the way shown in Figure 8.6(b), and measuring the deflection at point X. The model can be made of a different material from the prototype, as long as both are linear. We saw in Chapter 7 that the forces in redundant beams of this kind depend only on the ratios of the flexural rigidities of different parts of the beams. If the beam is not uniform, these ratios have to be the same in the model as in the prototype, and so of course must be the ratios of the span lengths, but otherwise the prototype does not need to be modelled precisely to scale. If the prototype is a uniform steel I-beam, for instance, the model need not be an I-beam, but can be a uniform strip of brass or plastic.

The reactions at supports 2 and 3 can be found in precisely the same way, by imposing appropriate displacements at 2 and 3 on an unloaded model of the beam.

This method can be applied to any linear structure. In frame structures the effects of axial forces in the members are generally small compared with those of bending moments, as we saw in Section 7.3, and it is then only necessary so to proportion the model that the flexural rigidities of different parts have the correct ratios. The deflections applied to the models can be quite large, though this may introduce small errors due to geometry change effects. Sometimes, though not in continuous beams, these errors can be eliminated or reduced by deflecting the model first in one direction and then an equal distance in the opposite direction, and measuring between the extreme positions.

Internal reactions within a structure can also be found. Suppose, for instance, that we wish to find the bending moment set up at point Y in Figure 8.6(a) by a unit load at point X. The displacement corresponding to bending moment at Y is a relative rotation between segments of the beam on either side of Y. Imagine, therefore, that a relative rotation ϕ takes place at Y (Figure 8.7(a)). As a result the beam deflects, but it remains in contact with its supports, and reactions are induced at those supports; call the corresponding loading (3). In the analysis it is easiest to think of a structure cut at Y, and of the moment and shear force at Y as applied externally. Figure 8.7(b) shows loading (1), the unit load at X we are interested in, as in Figure

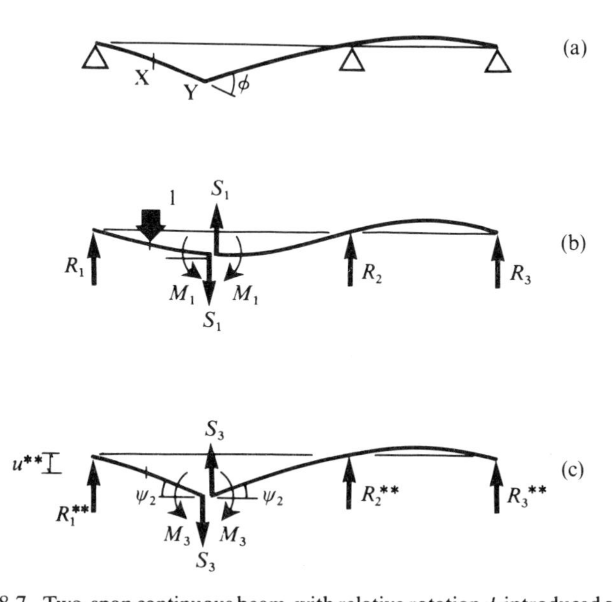

FIG. 8.7. Two-span continuous beam, with relative rotation ϕ introduced at Y.

8.6(a), except that now that the shear force and bending moment at an imaginary cut at Y have been added. Figure 8.7(c) shows loading (3) produced by a relative rotation at Y. The loads and their corresponding displacements for the two loadings are listed in Table 8.4.

Applying the reciprocal theorem to loadings (1) and (3):

$$u^{**}+S_1u_Y^{**}-M_1\psi_2-S_1u_Y^{**}-M_1\psi_3=S_3u_Y-M_3\psi_1-S_3u_Y+M_3\psi_1, \tag{8.4.8}$$

TABLE 8.4

Point	System (1)		System (3)	
	Force	Corresponding displacement	Force	Corresponding displacement
1	R_1	0	R_1^{**}	0
X	-1	$-u$	0	$-u^{**}$
Y (left side)	$-S_1$	$-u_Y$	$-S_3$	$-u_Y^{**}$
Y (left side)	$-M_1$	ψ_1	$-M_3$	ψ_2
Y (right side)	S_1	$-u_Y$	S_3	$-u_Y^{**}$
Y (right side)	M_1	ψ_1	M_3	$-\psi_3$
2	R_2	0	R_2^{**}	0
3	R_3	0	R_3^{**}	0

that is

$$u^{**} = M_1(\psi_2 + \psi_3)$$
$$= M_1\phi, \tag{8.4.9}$$

since $\psi_2 + \psi_3$ is the relative rotation at Y in loading (3), and therefore

$$M_1 = \frac{u^{**}}{\phi}, \tag{8.4.10}$$

so that the moment at Y induced by unit load at X is the deflection at X in loading (3) divided by the imposed relative rotation ϕ. Experimentally, a rotation of this kind can be introduced if a cut is made in the model beam at Y.

8.5. Problems

1. A uniform continuous beam rests on three simple supports, one at each end and one at the centre, so that the two spans are equal. Its moment–curvature relation is linear. It carries various distributed loads. One of the end supports then sinks by a distance Δ; the other two supports do not move. As a result the reaction exerted by the end support alters by R. Use symmetry and superposition to show that the same change in end reaction would occur if both end supports sank through $\Delta/2$, the centre remaining fixed. Thence or otherwise find R as a function of Δ.

2. A symmetrical linear portal frame with hinged feet carries a horizontal load P at the upper left-hand corner. Use symmetry and superposition to prove that
   (i) the induced bending moment at the centre of the beam is zero, and
   (ii) the horizontal reactions at the stanchion feet are equal, and thence find the vertical reactions at the stanchion feet and the bending moments throughout the frame.

   Assume that the elongations produced by axial forces are negligible. What happens if they are not negligible?

3. Repeat the analysis of Problem 2 for a portal frame with fixed feet, loaded in the same way.

4. A uniform beam of length L and flexural rigidity F is simply supported at either end, and loaded by end moments C_1 and C_2. The rotation at end 1 can then be shown to be

$$\frac{1}{3}\frac{C_1 L}{F} - \frac{1}{6}\frac{C_2 L}{F}.$$

Use this result to determine
(i) the end rotation if a moment C_1 is applied to a beam built-in at the far end,
(ii) the end rotation if a moment C_1 is applied to the end of a two-span continuous beam which rests on three simple supports, the two spans each having a length L,
(iii) the end rotation if a moment C_1 is applied to the end of a semi-infinite continuous beam, which rests on simple supports evenly spaced L apart.

5. Section 8.4 discusses the analysis of a linear two-span continuous beam. Consider a uniform beam with two equal spans, and the problem of finding the reaction at the left-hand support induced by unit load at the centre of the left-hand span.
(i) Using a model beam, determine the end reaction by the method described in Section 8.4. A thin metal or wooden spline can be used to represent the beam, and the supports can be drawing pins pressed into a drawing board. Measurements can be made with a ruler, or on a sheet of graph paper pinned to the board. Surprisingly accurate results can be obtained with very simple models, made from wire or even cardboard.
(ii) Alternatively, use the method of direct integration of $F\,\mathrm{d}^2v/\mathrm{d}x^2$ to determine the deflected shape of the beam under the loading of Figure 8.6(b), and then use eqn (8.4.7) to determine the end reaction. Compare it with your experimental result.
(iii) Alternatively, solve the problem by virtual work. Compare the different methods.

6. How would you solve Example 6 in Section 7.7 (page 146) by the indirect model method? Make a model, and compare your solution with the calculated one. What would you do if the load were distributed instead of concentrated?

7. The elastic frame shown in Figure 8.8 is of non-uniform cross-section, but is symmetrical about the line CF. The points B and D are pin-jointed to a rigid tie-rod.

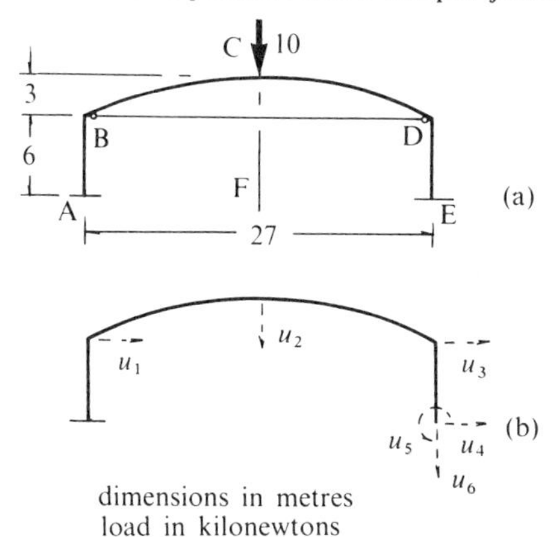

FIG. 8.8.

Figure 8.8(b) shows schematically a one-fiftieth scale model of the frame. In the model the flexural rigidities are reproduced to scale, but the tie-rod is omitted. Provision is made for applying and measuring displacements at the points and in the directions indicated. The results of three such tests are given in Table 8.5.

TABLE 8.5

	Imposed displacements					Measured displacements		
Test no.	u_1 mm	u_3 mm	u_4 mm	u_5 rad	u_6 mm	u_1 mm	u_2 mm	u_3 mm
1	+10	−10	0	0	0	—	−12·4	—
2	—	—	+10	0	0	+1·0	+7·0	+5·0
3	—	—	0	0·10	0	−2·6	−9·6	−5·4

For the full-scale frame, find the tension in the tie-rod and the reactions and bending moment at the support E due to the application of a 10 kN vertical load at C.

9. Plastic theory

9.1. Introduction

WHEN the applied loads are small, most engineering structures respond in the linear manner described in the previous chapter. This linear response reflects a linear relationship between strain and stress in the material from which the structure is made, and the fact that deflections are small. At larger stress levels the relationship between strain and stress is more complicated. Many different kinds of behaviour are observed, and this is an important subject of study and research in its own right, beyond the scope of the present book and dealt with in texts on solid mechanics, materials science, and metallurgy. There is however one particular kind of nonlinear behaviour, due to material properties, which very often arises in structural engineering practice, and has an important theory of its own, which this chapter introduces. A second kind of nonlinearity arises when the deflections under load significantly alter the geometry, and can occur even though the material from which the structure is made is linear: that is described in Chapter 10.

Tests on many structural materials show relationships between stress and strain† like the one shown in Figure 9.1. At low stresses the relation is linear,

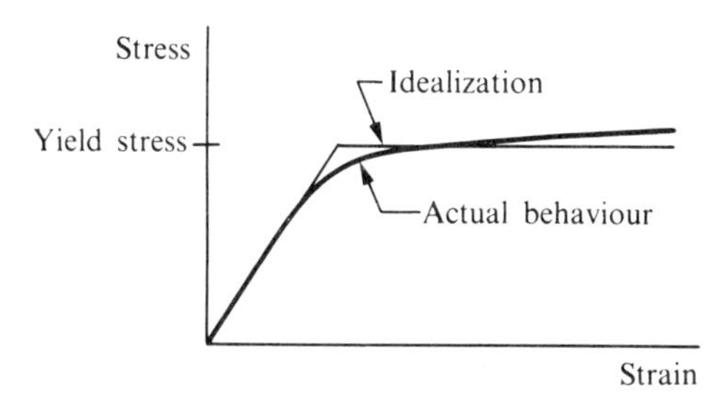

FIG. 9.1. Relationship between stress and strain for typical ductile material.

but at some larger stress level the material begins to yield, and the strain then increases more and more rapidly with increasing stress, so that the curve plotting the stress–strain relation bends over. In some materials, such as low-alloy steel, the transition is very sudden, and the stress–strain curve

† Complete definitions of the concepts of stress and strain will be found in books on solid mechanics. In outline, stress describes the forces exerted within a solid body, by one part on another, and strain describes its deformation. Here we only need an incomplete elementary definition: the stress in an axially-loaded uniform bar is its tension divided by its cross-sectional area, and its strain is the elongation divided by the length of the bar.

becomes almost flat, so that the strain can continue to increase to very large values at almost constant stress. In other materials, such as structural aluminium and titanium alloys, stainless steel, and steel which has been strengthened by alloying and heat treatment, the transition is more gradual, and the stress continues to increase with increasing strain, although at a much reduced rate.

It is useful to have an idealization which represents this kind of behaviour in a mathematically simple way. One simple idealization is included in Figure 9.1. The strain is supposed to increase linearly with the stress, up to a certain yield stress. At that yield stress, the strain can increase indefinitely without any further increase in stress. This idealized material is called an *elastic perfectly-plastic* material: *elastic* because to start with it behaves like a linear elastic material, *perfectly plastic* because once it has begun to yield plastically it does so at a constant stress. In the remainder of this chapter we shall be concerned with the mechanics of structures made from elastic perfectly-plastic materials. It is a useful idealization for many materials: for steel, for ductile metals generally, and sometimes even for reinforced concrete, clay, and ice. You will notice that the idealization says nothing about fracture, and implies that at the yield stress the strain can go on and on increasing without limit, and for this reason the idealization is most useful for materials that can deform extensively before fracture occurs. If fracture is important, as in reinforced concrete, the theory must be used cautiously.

Imagine now that a structural element, such as a beam or a bar in a framework, is made from a material which can be idealized as elastic perfectly-plastic. Exactly as there is a relationship between stress and strain, so there is between a stress resultant (such as bending moment in a beam or tension in a bar) and its corresponding deformation (such as curvature or elongation). The relationship is of the general form shown in Figure 9.2, which plots bending moment against curvature for a typical beam. When the bending moment is small, the curvature and the moment are linearly related.

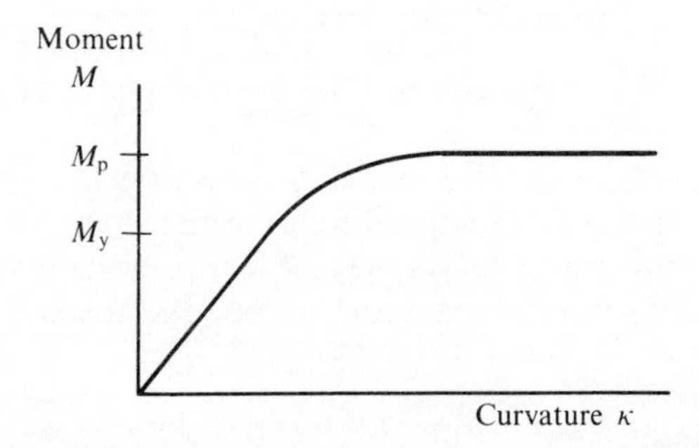

FIG. 9.2. Relationship between curvature and bending moment, for a beam made of an elastic–plastic material.

Beyond a *yield moment* M_y the curvature begins to increase more rapidly. Later a *full plastic moment* M_p is reached, at which the curvature can increase indefinitely, without any further increase in moment. The relationship between moment and curvature is more complicated than that between stress and strain, because the strain is not uniformly distributed within the beam, and so different parts of it reach the yield stress at different stages of the loading history. The yield moment and the full plastic moment depend on the dimensions of the element cross-section, and on quantities describing its material behaviour, such as yield stress. The full plastic moment can be calculated (see, for example, Baker, Horne, and Heyman (1956)), or can be found experimentally. Similar relationships can be derived for other stress resultants and their corresponding deformations†.

Next imagine a complete structure made from elements each made from an elastic perfectly-plastic material. It carries some distribution of loads, and all the loads are increased together in proportion. A convenient deflection is chosen as a measure of the deformation of the structure. If we plot the load intensity against the deflection, we find that they are related in the way shown in Figure 9.3. Once again, the first part of the graph is linear, and

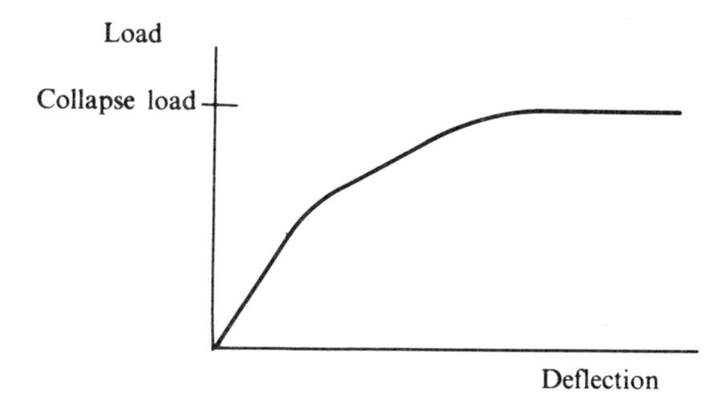

FIG. 9.3. Relationship between deflection and load, for a structure made from elements which are made from an elastic–plastic material.

reflects the fact that at low loads all the elements are still behaving linearly. Later, yield begins to occur somewhere in the structure, first at one point in one element, and later at more than one point in more than one element. Finally, a *collapse* state is reached, at which the loads cannot be increased any more and the deflections can go on growing without limit.

If an engineer has to know the details of the load–deflection relation for a particular elastic perfectly-plastic structure, he will follow the kind of

† Linear relationships in structural mechanics do not usually involve cross-effects between different kinds of stress resultants: thus the flexural rigidity of a beam, expressing a linear relation between moment and curvature, is independent of the tension in a beam. If plastic yield occurs, this is no longer always true: the presence of an axial tension will reduce the full plastic moment. However, this effect is often negligibly small (Baker and Heyman, 1969), and will be ignored here.

analysis described in Section 7.6, which, as we saw then, is fairly laborious, following step-by-step the development of plastic deformation through the structure. Much more commonly, however, the engineer is most keenly interested in the collapse load, because collapse is one of the limit states which constrain the loads that can be applied, but is not much interested in the details of deflections, though he may also calculate deflections under working loads (which are usually substantially smaller than the collapse load). It turns out that the collapse load can be found relatively easily, by an independent method which bypasses the need for a complete analysis. The next section explains how to do this by the methods of plastic theory.

Plastic theory is a relatively recent development in structural mechanics. Some of the ideas involved were thought about and experimented with in the early part of the century, but at the time few engineers made any use of them. The modern development of the subject dates from the Second World War, stimulated and led first by the work of Baker and his group at Cambridge in England, and later by research teams in the USA and elsewhere in Europe.

Most elastic perfectly-plastic skeletal structures have the kind of relationship between load and deflection shown in Figure 9.3. This is almost always so for beams and frames, and most of the theory has been developed for them, and will be explained in that context here. Plastic theory can also sometimes be applied to frameworks, but plastic instability in compression can there lead to structural element behaviour markedly unlike that in Figure 9.2, and thence to load–deflection curves which reach a peak and then fall rapidly. Arches, plates, and shell structures often behave in a similar way: the reason is that the deflections that occur as the structure begins to collapse significantly alter its geometry, and alter the balance of forces within it. Simple plastic theory can then give very misleading results.

9.2. Plastic collapse

Consider a simply supported uniform beam which carries a uniformly distributed load (Figure 9.4(a)). Its moment–curvature relation is the one shown in Figure 9.2. The intensity w of the load is steadily increased from zero. The beam is statically determinate, and the bending moment at x from the left-hand support is

$$M=\tfrac{1}{2}wx(L-x), \tag{9.2.1}$$

and its greatest value is $wL^2/8$ at the centre.

At first the moment is small enough for the curvature to depend linearly on the moment everywhere in the beam; the distribution of moment is as in Figure 9.4(b). Later the moment in a central zone exceeds the yield moment M_y (Figure 9.4(c)), and in that zone the curvature increases more rapidly. Finally the greatest moment reaches the full plastic moment M_p (Figure

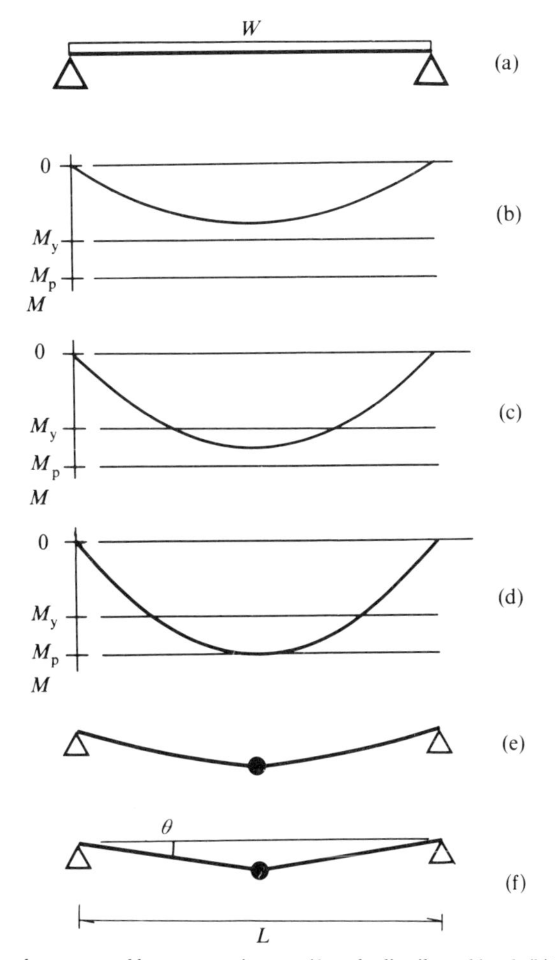

FIG. 9.4. (a) Simply supported beam, carrying a uniformly distributed load. (b) Distribution of bending moment, when maximum moment is less then M_y. (c) Distribution of bending moment, when maximum moment is more than M_y but less than M_p. (d) Distribution of bending moment, when maximum moment is M_p. (e) Deformed configuration when a plastic hinge has formed at the centre. (f) Incremental deformation during collapse.

9.4(d)), when

$$M_p = wL^2/8$$
$$w = 8M_p/L^2. \qquad (9.2.2)$$

Once that has happened, the load cannot be increased any further, and the curvature cannot change except at the very centre, where the bending moment is M_p and allows an indefinite increase in curvature. Any further deformation takes place at a plastic hinge at the centre of the beam, a point at which finite relative rotations can occur between segments on either side of the hinge, so that the deformed configuration will be that shown in Figure 9.4(e).

The beam can continue to collapse in this mode, under the collapse load intensity $8M_p/L^2$, all the deformation now being concentrated in the hinge. The deformation that occurs in an increment of collapse is that shown in Figure 9.4(f), where the straight segments on either side of the central plastic hinge indicate that once a collapse mechanism has been reached all the deformation occurs at the hinge, and none elsewhere. Plastic hinges are indicated in diagrams by solid circles, which distinguishes them from 'free' hinges indicated by open circles.

As collapse proceeds, the load descends, so that it does work on the beam, and further rotation occurs in the hinge, where work is dissipated plastically in producing relative rotation against the full plastic moment. Another way of finding the collapse load is to make use of a work balance between the work done and the work dissipated. Suppose that the beam moves incrementally in the collapse mechanism so that the left-hand segment rotates clockwise through a small angle θ about its left-hand support, and the right-hand segment rotates counter-clockwise through θ (Figure 9.4(f)). The relative rotation at the hinge is 2θ, from geometry, and so the work dissipated there is $M_p(2\theta)$. The left-hand segment rotates about its left end, which does not move vertically, but at the centre the vertical displacement is $L\theta/2$ downward. Vertical displacement is proportional to the distance from one end, and so the mean vertical displacement is $L\theta/4$. The uniformly distributed load on a segment is $wL/2$, and so the work done by the load on the segment as the beam moves is $(wL/2)(L\theta/4)$. The same amount of work is done by the load on the other segment, and so the total work done by the load is $wL^2\theta/4$. Equating the work done to the work dissipated,

$$wL^2\theta/4 = 2M_p\theta, \tag{9.2.3}$$

and so at collapse

$$w = 8M_p/L^2. \tag{9.2.4}$$

This very simple example illustrates several features of plastic collapse which apply generally. Firstly, collapse occurs by the formation of plastic hinges. Once enough hinges have formed to make a collapse mechanism, no further deformation occurs outside the hinges. The collapse load can be got either by an approach based on statics, as in Figure 9.4(d), or by a work calculation based on a collapse mechanism (Figure 9.4(f)), eqn (9.2.3)).

Statically indeterminate beams behave in the same way. Figure 9.5 describes the behaviour of a uniform beam built-in at both ends, again loaded with a uniformly distributed load which increases from zero. It shows

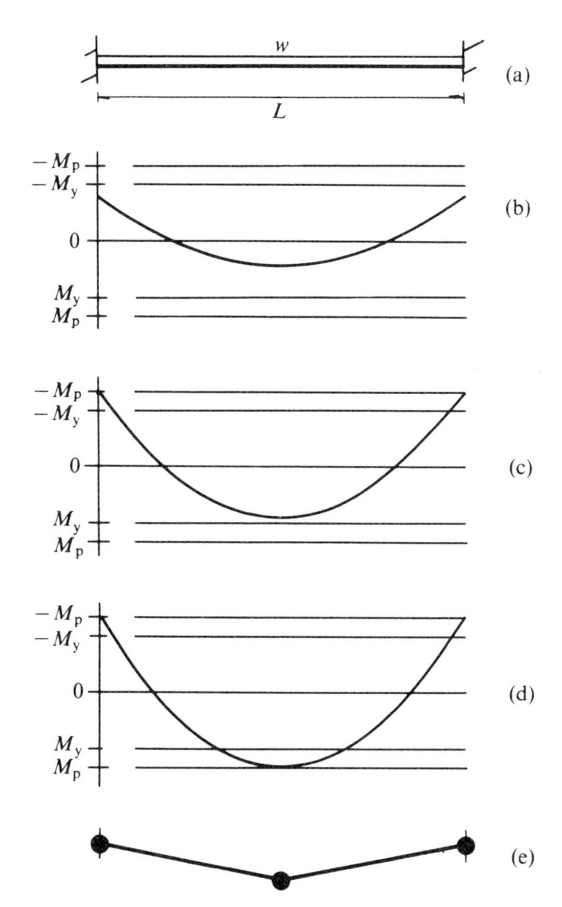

FIG. 9.5. (a) Beam built-in at both ends, carrying a uniformly distributed load. (b) Distribution of bending moment, when maximum moment is less than M_y. (c) Distribution of bending moment, when moment at the built-in ends is $-M_p$, but that at the centre is less than M_y. (d) Distribution of bending moment, when plastic hinges have formed at either end, and at the centre. (e) Incremental deformation during collapse.

in turn the distribution of bending moment when the beam is still wholly elastic (Figure 9.5(b)), when plastic hinges have formed at the ends (Figure 9.5(c)), and when plastic hinges have formed at the ends and at the centre (Figure 9.5(d)), and a collapse mechanism (Figure 9.5(e)) has been reached. A third example of this behaviour was analysed in Example 7.6.1 (page 140), which followed the elastic–plastic analysis through in detail.

In the first example above the statical and mechanism methods were used to find the exact collapse load. The statical method can also be used to find a lower bound on the collapse load, a value which is known to be less than or equal to the exact value. The mechanism approach can be used to find an upper bound, a value which is known to be greater than or equal to the collapse value. Together they bracket the exact collapse load. If the two bounds coincide, they must do so at the exact value. The fact that the methods give lower and upper bounds is a consequence of two limit theorems, which are extremely useful in plastic theory.

9.3. Limit theorems

The theorems that follow are phrased in terms of the plastic collapse of beams and frames. They can readily be generalized.

Upper-bound theorem. If an estimate of the plastic collapse load of a structure is made by equating the internal dissipation of energy to the work done by the external loads, in any postulated mechanism of deformation of the structure, the estimate will be either high, or correct.

Lower-bound theorem. If any distribution of bending moment in the structure can be found, which is everywhere in equilibrium internally and balances certain external loads, and at the same time does not anywhere exceed the full plastic moment, those loads will be carried safely by the structure, which will not collapse.

Alternative statements of the theorems are possible: these are based on generalized formulations by Calladine (1969). The lower-bound theorem is proved in Appendix I, and the proof of the upper-bound theorem is similar.

The use of the theorems is best illustrated by an example.

Example 9.3.1. A uniform beam of length L is simply supported at one end and built-in at the other end (Figure 9.6(a)). It carries a uniformly distributed load. Find upper and lower bounds on the load intensity which will produce plastic collapse.

A complete elastic–plastic analysis of this problem was carried out in Example 7.6.1. Here we are only concerned with the collapse load intensity, and not with the detailed development of deflection.

It is usually easiest to use the upper–bound theorem first. The beam is not symmetric, and so symmetry alone cannot tell us where the plastic hinges are going to be. We might guess that one plastic hinge will form near the centre (Figure 9.6(b)). That, however, is not a collapse mechanism: the segment to the left of the hinge cannot move downwards by rotation about the simple support, because the hinge is supported by the right-hand segment acting as a cantilever. A collapse mechanism needs two hinges, one at the right-hand end and one somewhere near the centre, as in Figure 9.6(c), and movement by rotation can then occur. It would still be a perfectly good mechanism if the right-hand hinge were away from the end, but it seems unlikely that such a mechanism would be the most critical one, if only because that would imply that the collapse load was independent of the beam length. It can easily be shown more rigorously that in the actual collapse mechanism the right-hand hinge must be at the right-hand end.

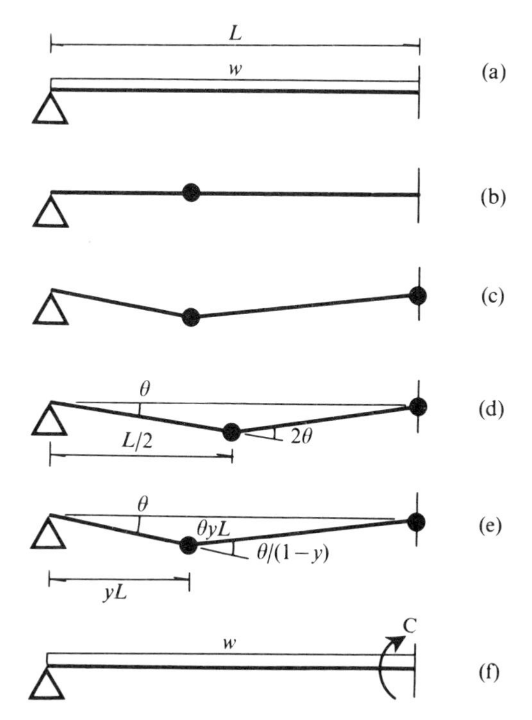

FIG. 9.6. (a) Beam simply supported at one end and built-in at the other end, carrying a uniformly distributed load. (b) One plastic hinge, near to the centre of the beam. (c) Collapse mechanism, formed by two plastic hinges, one at the built-in end, the other near to the centre. (d) Collapse mechanism, formed by two plastic hinges, one at the built-in end and the other at the centre. (e) Collapse mechanism, formed by two plastic hinges, one at the built-in end and the other at a distance yL from the other end. (f) Redundant support moment C at the built-in end.

Suppose the left hinge to be at the centre (Figure 9.6(d)), and the segment to the left of it to rotate θ clockwise about the simple support. The hinge moves downwards by $L\theta/2$, and the right-hand segment rotates counter-clockwise by

$$\frac{L\theta/2}{L/2} = \theta.$$

The relative rotation at the right-hand hinge is θ, and that at the centre is 2θ. The mean vertical displacement on each segment is $L\theta/4$, and so, since the load is uniformly distributed, the work done by the load is $(wL)(L\theta/4)$. So, making an estimate of the collapse load by the method suggested in the upper-bound theorem,

$$\underset{\substack{\text{work done}\\ \text{by load}}}{wL^2\theta/4} = \underset{\substack{\text{central}\\ \text{hinge}}}{(M_p)(2\theta)} + \underset{\substack{\text{end}\\ \text{hinge}}}{(M_p)(\theta)}, \tag{9.3.1}$$

whence

$$w = 12M_p/L^2. \tag{9.3.2}$$

The theorem tells us that this is certainly greater than or equal to the collapse load intensity.

The beam is not symmetric, and so there is no reason why the plastic hinge should be at the centre. We now investigate a more general mechanism (Figure 9.6(e)), in which the left hinge is at a distance yL from the simple support, the parameter y not yet being given a particular value. The mechanism in Figure 9.6(d) corresponds to $y = \frac{1}{2}$.

Let the left-hand segment again rotate through θ. The hinge moves downward by θyL, and the right-hand segment rotates through

$$\frac{\theta yL}{(1-y)L} = \frac{\theta y}{1-y},$$

which is the rotation of the right-hand hinge. The relative rotation at the other hinge is

$$\theta + \frac{\theta y}{1-y} = \frac{\theta.}{1-y},$$

recalling elementary geometry (the exterior angle of a triangle is the sum of the interior angles at the other two vertices).

The work balance is then

$$\underset{\text{work done}}{(wL)(\tfrac{1}{2}\theta yL)} = \underset{\substack{\text{hinge}\\ \text{at } yL}}{M_p\frac{\theta}{1-y}} + \underset{\substack{\text{hinge at}\\ \text{built-in}\\ \text{end}}}{M_p\frac{\theta y}{1-y}} \tag{9.3.3}$$

$$= M_p\theta\frac{1+y}{1-y}, \tag{9.3.4}$$

and so

$$w = 2\frac{1+y}{y(1-y)}M_p/L^2. \tag{9.3.5}$$

This is an upper bound on the collapse load intensity for any value of y (within the range 0 to 1, of course). The most useful upper bound is the smallest one, and so the expression for w should be minimized with respect to the free parameter y, which locates the left plastic hinge. This can be done by elementary calculus. The minimizing value is

$$y = \sqrt{2} - 1, \tag{9.3.6}$$

and then

$$w = (6+4\sqrt{2})M_p/L^2$$
$$= 11{\cdot}66\ M_p/L^2. \qquad (9.3.7)$$

Close to the minimizing value the bound is very insensitive to y. If we guess y to be 0·45, the corresponding w is 11·72 M_p/L^2; if we guess it to be 0·4, w is 11·67 M_p/L^2.

Now attack the same problem with the lower-bound theorem. If the fixing moment at the right-hand support is C (Figure 9.6(f)), the bending moment at x from the simple support is

$$M = \tfrac{1}{2}wx(L-x) - Cx/L. \qquad (9.3.8)$$

The maximum value of M with respect to x can be found by calculus, and is

$$wL^2/8 - \tfrac{1}{2}C + \tfrac{1}{2}C^2/wL^2.$$

Its minimum value is at the built-in end $x = L$, and is $-C$. Everywhere else the bending moment lies between these two extremes. The conditions of the lower-bound theorem require the full plastic moment not to be exceeded anywhere, and so both extreme values must lie within the range $-M_p$ to $+M_p$, or, equivalently, the absolute value of neither extreme value should exceed M_p. It follows that the theorem requires that both

$$C \leqslant M_p \qquad (9.3.9)$$

$$wL^2/8 - \tfrac{1}{2}C + \tfrac{1}{2}C^2/wL^2 \leqslant M_p. \qquad (9.3.10)$$

If, for example, C is zero, the first condition is satisfied automatically, and the second if

$$wL^2/8 \leqslant M_p, \qquad (9.3.11)$$

that is, if w is no greater than $8M_p/L^2$, which is therefore a lower bound on the collapse value of w.

If C is $wL^2/10$, say, the first condition is satisfied if

$$w \leqslant 10wL^2/M_p \qquad (9.3.12)$$

and the second if

$$wL^2(\tfrac{1}{8} - \tfrac{1}{20} + \tfrac{1}{200}) \leqslant M_p, \qquad (9.3.13)$$

that is, if

$$w \leqslant 12{\cdot}5\ M_p/L^2. \qquad (9.3.14)$$

The first condition is now the more stringent, and so the conditions of the theorem are met if

$$w \not> 10\ M_p/L^2, \qquad (9.3.15)$$

which is an improved lower bound on the collapse value of w.

Any value of C gives a lower bound. The most useful lower bound is the largest one. The second condition, eqn (9.3.10), is the more critical if C is small, and the first, eqn (9.3.9), is the more critical if C is large. The largest lower bound is obtained if both conditions at once are just satisfied, so that

$$C = M_p \tag{9.3.16}$$

and

$$wL^2/8 - \tfrac{1}{2}C + \tfrac{1}{2}C^2/wL^2 = M_p. \tag{9.3.17}$$

Eliminating C, and solving the resulting quadratic equation for w

$$w = (6 + 4\sqrt{2})M_p/L^2, \tag{9.3.18}$$

which coincides with the upper bound found independently by the work method. Since both upper and lower bounds coincide, this must be the exact collapse load intensity.

9.4. Portal frames

Example 9.4.1. A portal frame with fixed stanchion feet (Figure 9.7) is made from a section with a uniform full plastic moment M_p. It carries a horizontal load U, at the top of one stanchion, and a vertical load V at the centre of the beam. What values can U and V have if the frame is not to collapse?

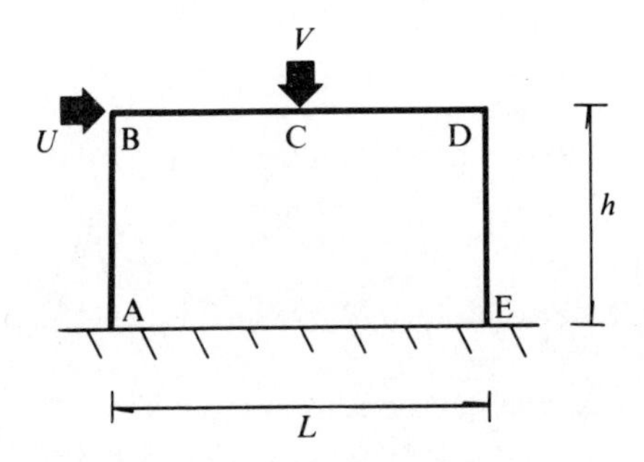

FIG. 9.7. Portal frame.

It is easiest to approach this problem by the upper-bound collapse-mechanism technique, but to use statics to help one to decide where the plastic hinges are going to be. Plastic hinges will form at points where the bending moment has extreme values. These will either be at the ends of elements, as at the right-hand hinge in Figure 9.6(e), or at stationary values where the moment has a maximum or a minimum, as at the centre of the beam in Figure 9.4, or at points where concentrated loads act. It was shown in Section 4.2 that the derivative of moment with respect to distance along a straight element is the shear force (eqn (4.2.4)), and so a stationary value of the moment corresponds to zero shear force. In this problem, since no lateral loads act between A and B, between B and C, between C and D, or between D and E, the shear force must be uniform within each of these segments, and so the bending moment can only have an extreme value at one of the points A, B, C, D, and E. It follows that we need only consider mechanisms that have hinges in these five positions.

There turn out to be three distinct collapse mechanisms, illustrated in Figure 9.8, where the mechanisms are on the left and the corresponding

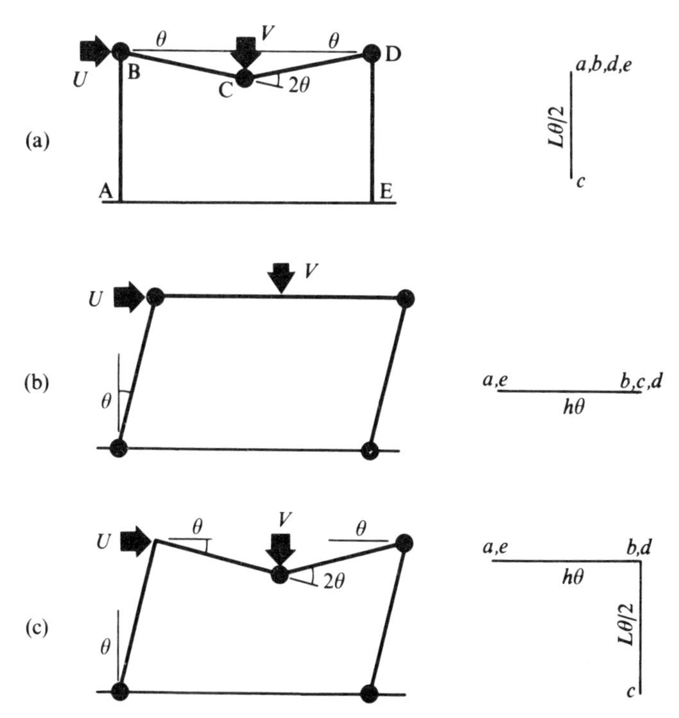

FIG. 9.8. (a) Collapse mechanism I, and corresponding displacement diagram. (b) Collapse mechanism II, and corresponding displacement diagram. (c) Collapse mechanism III, and corresponding displacement diagram.

displacement diagrams are on the right. In each mechanism θ is supposed to be an infinitesimally small angle, corresponding to an increment in the collapse mechanism, but in the diagram it is exaggerated. Deformations by plastic hinge formation are purely in bending, without any axial elongation, and displacement diagrams help one to check that this condition is being obeyed.

In mechanism I the beam alone collapses, behaving as did the beam with built-in ends in Figure 9.5; this is a mechanism which could still occur even if the stanchions were completely rigid. If the left-hand half of the beam rotates through θ about B, the central hinge at C descends through $L\theta/2$. The horizontal load does no work, because its point of application does not move. The work equation is

$$\begin{array}{ccccccc} V(L\theta/2) & = & M_p\theta & + & M_p(2\theta) & + & M_p\theta, \\ \text{work} & & \text{hinge} & & \text{hinge} & & \text{hinge} \\ \text{done} & & \text{at B} & & \text{at C} & & \text{at D} \end{array} \tag{9.4.1}$$

whence

$$VL/M_p = 8. \tag{9.4.2}$$

If this condition is satisfied, collapse by mechanism I can occur. If VL/M_p is less than 8, collapse by mechanism I will not occur, because the load V cannot put in enough work to overcome the plastic resistance of the hinges in that mechanism, but it may be that the frame can still collapse by some other mechanism.

In mechanism II the frame sways sideways, deforming from a rectangle into a parallelogram by hinge formation at the corners A, B, D, and E. The vertical load moves only sideways, and does no work. The work calculation is

$$\begin{array}{ccccccccc} U(h\theta) & = & M_p\theta & + & M_p\theta & + & M_p\theta & + & M_p\theta, \\ & & \text{hinge} & & \text{hinge} & & \text{hinge} & & \text{hinge} \\ & & \text{at A} & & \text{at B} & & \text{at D} & & \text{at E} \end{array} \tag{9.4.3}$$

whence

$$Uh/M_p = 4. \tag{9.4.4}$$

Mechanism III is a combined mechanism, in which ABC rotates through θ as a rigid body, the angle of B remaining a right angle, and B moving horizontally through $h\theta$. Since BCD does not alter in length points B, C, and D must move through equal distance horizontally. Segment DE rotates clockwise through θ, so that D shall also move horizontally through $h\theta$, and CD rotates counter-clockwise through θ since otherwise the vertical displacements at C would not match. Relative rotations are θ at A, zero at B, 2θ at C, 2θ at D, and θ at E. The work equation is

$$\begin{array}{ccccccccc} U(h\theta) + V(L\theta/2) & = & M_p\theta & + & M_p(2\theta) & + & M_p(2\theta) & + & M_p\theta, \\ & & \text{hinge} & & \text{hinge} & & \text{hinge} & & \text{hinge} \\ & & \text{at A} & & \text{at C} & & \text{at D} & & \text{at E} \end{array} \tag{9.4.5}$$

and so

$$Uh/M_p + \tfrac{1}{2}(VL/M_p) = 6. \tag{9.4.6}$$

Bringing the collapse conditions for the three mechanisms together, the frame will not collapse in any of the three modes if all the following conditions are satisfied

$$VL/M_p \leqslant 8, \tag{9.4.7}$$

$$Uh/M_p \leqslant 4, \tag{9.4.8}$$

$$Uh/M_p + \tfrac{1}{2}(VL/M_p) \leqslant 6. \tag{9.4.9}$$

A convenient way of representing these conditions is the graph in Figure 9.9, called in plastic theory an interaction diagram, which plots the non-dimensional vertical load VL/M_p on one axis and the non-dimensional

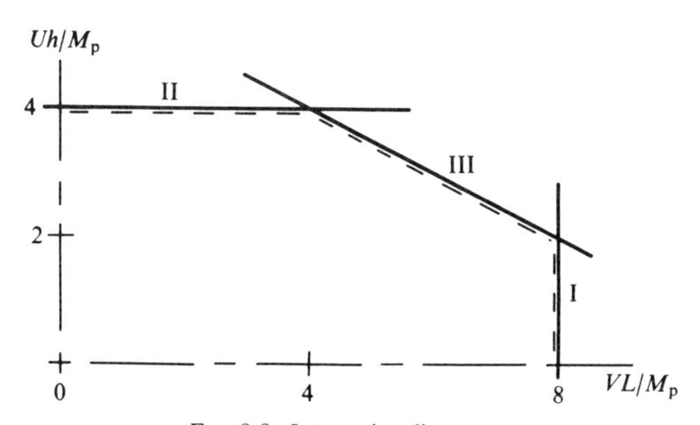

FIG. 9.9. Interaction diagram.

horizontal load Uh/M_p on the other. Any point on such a diagram represents a combination of vertical and horizontal load. Inequality (9.4.7), which expresses the condition that collapse does not occur by mechanism I, is obeyed to the left of the line marked I in the diagram. Similarly, inequality (9.4.8), for mechanism II, is obeyed below the line marked II, and inequality (9.4.9) is obeyed below and to the left of the line marked III. In the region marked out by the dashed boundary, all three conditions are met, and collapse will not occur under loads represented by points in this region. Loads represented by points on the dashed boundary can just produce collapse. Loads represented by points outside it cannot be carried.

This has been an upper–bound calculation. If we are to be sure that the actual collapse condition is given by the dashed boundary in Figure 9.9, we must either be certain that no possible fourth alternative mechanism has been overlooked, or the upper bound must be complemented by a lower-bound calculation, which must show that the loads can indeed lie on the collapse boundary and yet not break the conditions of the lower-bound theorem. It is possible to carry out a general lower-bound analysis of this portal frame, for any ratio of VL to Uh, but rather awkward to do so. The method of complementing the upper–bound calculation by a lower-bound is best illustrated by a specific numerical case, and this is done in Example 9.4.2. In fact there are no further more critical mechanisms, and the boundary in Figure 9.9 does locate the exact collapse loads.

Example 9.4.2. A portal frame with built-in stanchion feet is 10 m across and 4 m high (Figure 9.10(a)). Its beam is made from a section which has a full plastic moment of 250 kN m, and its stanchions from a section which has a full plastic moment of 210 kN m. It carries vertical and horizontal loads of 100 kN and 70 kN. By what factor can these loads be multiplied before collapse occurs?

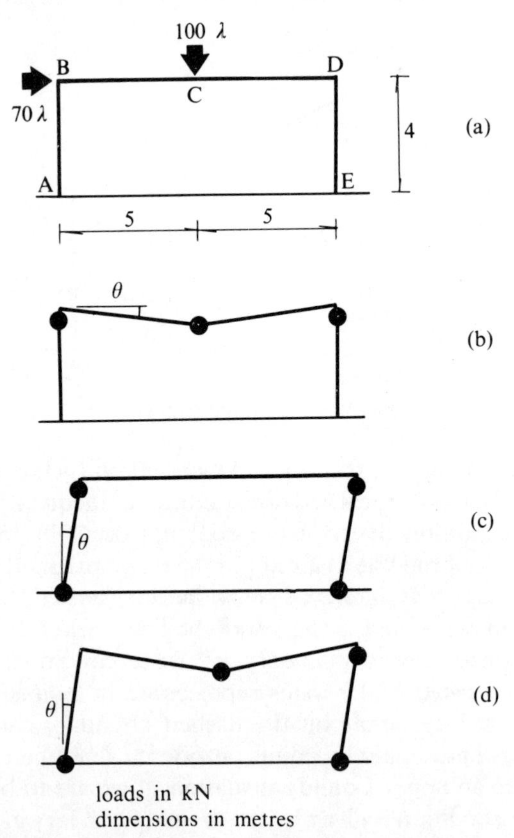

FIG. 9.10. (a) Portal frame. (b) Mechanism I. (c) Mechanism II. (d) Mechanism III.

This frame has one new feature, in that its section is not uniform, and the stanchions are weaker than the beam. At the corners B and D the moment is the same on both sides of the joint, otherwise the joint itself could not be in equilibrium. It follows that if there is a plastic hinge at B or D it must be in the stanchion rather than the beam, and care must be taken to use the appropriate full plastic moment. In mechanism diagrams this is indicated by drawing the solid circle close to the corner but in the section where the hinge will form, as in Figure 9.10(b).

The three mechanisms are shown in Figure 9.10(b), (c), and (d). In the upper–bound calculation we use the original loads of 100 and 70 kN multiplied by a load factor λ, and for each mechanism find the value that λ has to have if collapse by that mechanism is to occur. The work equations corresponding to the three mechanisms are:

MECHANISM I

$$(100\lambda)(5\theta) = 210\theta + 250(2\theta) + 210\theta$$

$$\lambda = 1{\cdot}84; \tag{9.4.10}$$

MECHANISM II

$$(70\lambda)(4\theta) = 210\theta + 210\theta + 210\theta + 210\theta$$

$$\lambda = 3; \tag{9.4.11}$$

MECHANISM III

$$(100\lambda)(5\theta) + (70\lambda)(4\theta) = 210\theta + 250(2\theta) + 210(2\theta) + 210\theta$$

$$\lambda = 1{\cdot}718. \tag{9.4.12}$$

It appears that mechanism III is the governing one, and that collapse will just occur if the loads are multiplied by 1·718, so that they become 171·8 kN and 120·3 kN. In order to check this by a lower–bound approach, we have to confirm that these loads can be in equilibrium with a distribution of bending moments which everywhere meets the conditions of the lower-bound theorem.

Suppose that collapse mechanism III is the right one. There are plastic hinges at A, C, D, and E, and we know the bending moments at these points, because we know that they must be equal in absolute value to the full plastic moments, and we know their signs because we know which way the hinges are bending. Figure 9.11(a) shows the frame divided into segments AB, BC, CD, and DE. The moments between the segments at C and D, and between the segments and the foundation at A and E, are included in the diagram. We do not yet know the moment at B, nor the forces between the segments or the reactions of the foundation, but we now find them by repeated use of the equilibrium conditions.

Segment DE is in equilibrium. Taking moments about D:

$$0 = 4Q_e - 210 - 210$$

and so

$$Q_e = 105 \text{ kN}. \tag{9.4.13}$$

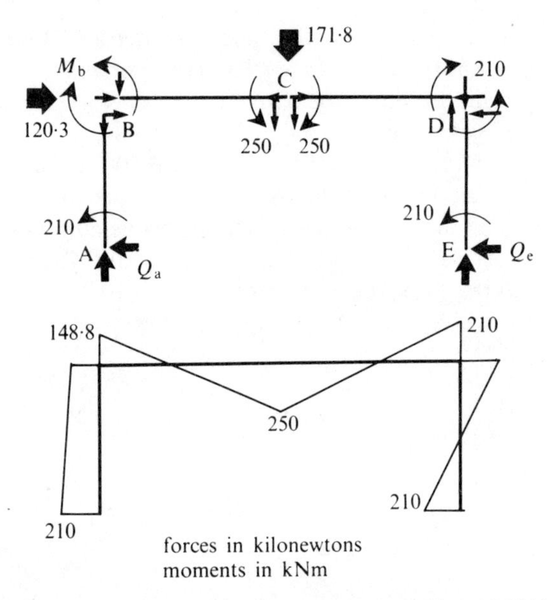

FIG. 9.11. (a) Forces and moments within the portal frame of Figure 9.10(a), at the collapse load corresponding to mechanism III. (b) Corresponding bending moment distribution.

The resultant horizontal force on the whole frame must be zero, and so

$$0 = 120{\cdot}3 - Q_a - Q_e$$

$$Q_a = 120{\cdot}3 - 105 = 15{\cdot}3 \text{ kN}. \tag{9.4.14}$$

Segment AB is in equilibrium. Taking moments about B

$$0 = 4Q_a + M_b - 210$$

and so

$$M_b = 210 - 61{\cdot}2 = 148{\cdot}8 \text{ kN m}. \tag{9.4.15}$$

We now know the moments at A, B, C, D, and E and know too that between each of these points and the next the moment varies linearly, because the shear force is constant. The bending moment diagram is Figure 9.11(b). Nowhere does the bending moment exceed the local full plastic moment, and so the conditions of the lower-bound theorem can be met when λ is 1·718, and that is the collapse load factor.

The above calculation took as its starting point the multiplied loads and the values of the moments as the assumed plastic hinge positions. One might be suspicious of a danger that this will lead to a circular argument, because of the assumption about where the hinges are. That suspicion is groundless: the lower-bound theorem allows any distribution of bending moment that obeys equilibrium and the full plastic moment condition, and the moments at A, C, D, and E serve simply as convenient starting values for the equilibrium calculations.

Imagine that we had overlooked mechanism III, and had supposed mechanism I to be the governing one, so that λ is 1·84 and the factored loads are 184 kN and 128·8 kN. An attempt to make a lower-bound check on this result follows the same lines as before. Again we imagine the frame divided into segments (Figure 9.12), and calculate forces and moments between

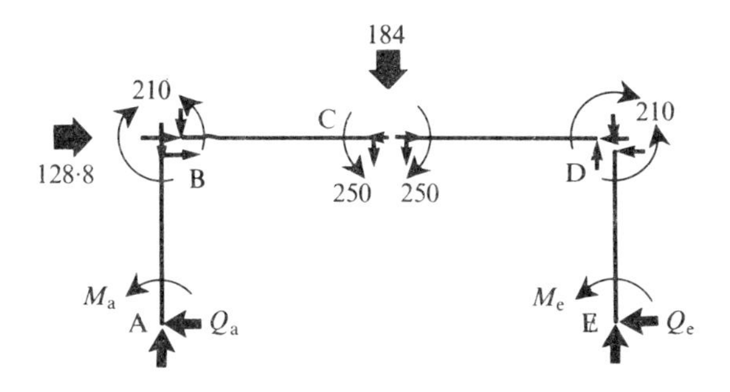

FIG. 9.12. Forces and moments within the portal frame of Figure 9.10(a), under the (erroneous) assumption that collapse is by mechanism I.

them, starting from the assumption that the full plastic moment is reached at B, C, and D. Since the whole frame is in equilibrium,

$$Q_a + Q_e = 128{\cdot}8. \tag{9.4.16}$$

Taking moments about B, for the equilibrium of AB

$$0 = 210 + 4Q_a - M_a. \tag{9.4.17}$$

Taking moments about D, for the equilibrium of DE

$$0 = -210 + 4Q_e - M_e. \tag{9.4.18}$$

Adding these last two equations

$$0 = 4(Q_a + Q_e) - M_a - M_e, \tag{9.4.19}$$

and using eqn (9.4.16)

$$M_a + M_e = 515{\cdot}2 \text{ kN m}. \tag{9.4.20}$$

If eqn (9.4.20) is to hold, at least one of M_a and M_e must be greater than 210 kN m, which is the local full plastic moment at both A and E. This breaks the conditions of the lower-bound theorem, and tells us that a load factor of 1·84 is only an upper bound, and not a correct value, and that we must look further for the true collapse mechanism. It suggests that we should look for mechanisms which have at least one hinge at A or E.

In this instance only the sum of the moments M_a and M_e could be found by statics, but not the individual values. That happens quite often when collapse mechanisms do not involve the complete structure, so that not all of it is statically determinate in the collapse state. Frequently the simple methods used above are all that we need to carry out a check that the lower-bound conditions are obeyed, but there are more systematic methods, described in texts on plastic theory (Heyman 1971).

9.5. Problems

In each of the following problems you should use both a mechanism upper–bound method and a statical lower-bound method to determine collapse loads. Except in Problem 13, and possibly in Problems 4 and 7, you should find it straightforward to determine coincident upper and lower bounds.

1. A uniform simply supported elastic–plastic beam of length L carries a central concentrated load. At what value of the load does collapse occur?

2. A uniform simply supported elastic–plastic beam of length L carries a concentrated load at a distance yL from one end. At what value of the load does collapse occur?

3. A uniform beam is built-in at one end and simply supported at the other, and carries a concentrated load at yL from the simply supported end. At what value of the load does plastic collapse occur?

4. A uniform cantilever is built-in at one end, and supported in addition by a simple support distant zL from the other end, where $z < \frac{1}{2}$. A vertical load W can act either at the free end or at a point $L/2$ from each end. For an arbitrary position of the support, find the value of the full plastic moment M_p necessary if collapse is not to occur for either position of the load. Thence find the optimum position of the support, which minimizes M_p, and the corresponding value of M_p.

5. A uniform simply supported elastic–plastic beam 8 m long carries a distributed load whose intensity varies linearly from 30 kN m at one end to zero at the other. The full plastic moment of the cross-section is 200 kN m. What is the collapse load factor, the factor by which the loading intensity has to be multiplied for collapse to take place?

6. An elastic–plastic continuous beam $3L$ long rests on three simple supports, unequally spaced, so that one span is $2L$ and the other L. The longer span has a full plastic moment $2M_0$, and the shorter span a full plastic moment M_0. The longer span carries a concentrated load at its centre. At what value of this load will collapse occur?

7. The beam in Problem 6 now carries instead a uniformly distributed load of intensity w, which extends over the whole beam. At what value of w will collapse occur, and what are then the three support reactions?

8. A vertical riser is fixed to one of the legs of an offshore oil production platform. It consists of a uniform continuous pipe, held by clamps at 12 m spacing, and free

between the clamps. A supply boat collides with the riser midway between two clamps. After the accident it is found that the riser is no longer straight, but at the impact point has been deflected 400 mm from its original line between the clamps, which have not moved. Assuming that the riser can be treated as an elastic–plastic beam with full plastic moment 100 kN m, estimate

(i) the force that must have been applied by the supply boat,
(ii) the minimum velocity it can have had at the instant of impact, if its mass is 2×10^5 kg.

9. Return to the lower-bound statical check in Example 9.4.2. In that analysis we did not find all the forces within the framework at collapse, but only those that were needed to find the bending moment distribution. Continue the analysis, and find

(i) the axial force in the beam,
(ii) the shear force distribution in the beam,
(iii) the vertical reactions at A and E.

10. A rectangular uniform elastic–plastic portal frame is like the frame studied in Example 9.4.1 (Figure 9.7, page 174), except that both stanchion feet are pinned to the foundation by free hinges, rather than built-in. It carries a horizontal load U and a vertical load V, placed as in Example 9.4.1. Construct the interaction diagram for this frame, and make statical checks on the calculated collapse loads for the following cases (a) $U = 0$ (b) $V = 0$ (c) $Uh = VL/2$.

11. A rectangular portal frame is 6 m in width and 3 m in height. The full plastic moment is 60 kN m for the beam and 40 kN m for each of the stanchions. The stanchion feet are built-in. A horizontal load of 30 kN acts at one corner, and a vertical load of 40 kN acts on the beam, not at the centre but at 4 m from the corner loaded by the horizontal load. Find the collapse load factor.

12. Repeat Problem 10, but now let the vertical load be uniformly distributed along the beam. Begin by finding upper bounds from the same mechanisms as you used in Problem 10, and draw corresponding lines on an interaction diagram. Then consider a more general mechanism, and thence construct the complete interaction diagram.

13. A semi-circular arch of radius R is made from a section with uniform full plastic moment M_0. It is built-in at both abutments, which are at the same level. It carries a central concentrated load. At what value of this load will collapse occur?

10. Stability

10.1. Introduction

TAKE a long thin strip of wood or metal, ideally about 500 mm long and 2 mm by 2 mm in cross-section, although the precise dimensions are unimportant. Grasp the ends between your fingers, and pull them outwards, so that the strip is in tension. You will find that in tension the strip can carry quite a large force, and it is unlikely that you will be strong enough to break it or to make it yield plastically. Now reverse the direction of loading, and push the ends towards each other so that the strip is compressed. If you push gently, nothing unusual will happen, and the strip will carry a small compressive load. If you push harder, at quite a low load the strip will suddenly bow sideways. You will find that once this has happened the load cannot be increased: if you move your fingers closer together the strip bows further, but the load is about the same. In the end the strip will be so severely bent that it cracks in the middle, if it is made of wood, or, if it is metal, yields so that when the load is removed it stays bent. If you repeat the experiment with a shorter strip, half the length of the first one, you will find that the same thing happens, but at a much larger load. If you do it again, with a still shorter strip, a quarter the length of the first one, you will find that still larger loads can be applied, and it may be difficult to apply enough force with your fingers to produce the sideways bowing.

Now take a strip of writing paper, and roll it into a cylinder about 5 mm in diameter, and glue it so that it remains in that form. If you load this cylinder axially, you will find that it behaves in essentially the same way as the wooden strip: it will carry a large load in tension, but only a much smaller one in compression, and the compressive load is limited by sideways bowing. Finally, take a second sheet of paper, and make a larger glued cylinder, about 50 mm in diameter. You will find that its behaviour in compression is different. Instead of the whole cylinder bowing sideways, localized crinkles and folds form at one cross-section. Again, however, the load the cylinder can carry is limited: an attempt to make the cylinder carry more load leads to growth of the crinkles, and they spread further into the cylinder.

These are instances of the phenomenon we call *buckling*. We can see from the experiments that buckling is likely to be an important factor that limits the loads that structural elements can carry. Buckling of the strip was an example of the general buckling of a complete structural element, and it is this kind of buckling that we shall be mainly concerned with. The crinkling of the larger cylinder is an example of local buckling, which does not involve the whole element, though it just as surely limits its capacity to bear loads.

The distinction between local and general buckling is not always very precise, and sometimes they interact, as in the case of the smaller paper cylinder, where over-all bowing is accompanied by local crinkling. Local buckling is an important factor limiting the strength of thin-walled structures, such as box girders, tubular columns, aircraft wings, and externally-pressurized tanks.

Buckling is linked to a wide class of behaviour that is called *instability*, and is important not only in mechanics but in electronics, in control engineering, in fluid mechanics, and indeed in almost every kind of system. We divide systems into *stable* and *unstable* systems, according to how they respond to small disturbances. A stable system responds to a small disturbance by a small and strictly limited response, which does not much alter its state or the way it responds to future disturbances, but an unstable system responds by a large response, which can take it into quite a different state. Imagine, for instance, a sharpened pencil lying on a flat horizontal table. A small push moves the pencil a little, so that it slides or rolls, but the response is limited, and the smaller the original push was, the less the movement is. The pencil lying flat has no tendency to set off on its own as a consequence of the initial push. This is an example of stable equilibrium, in which a small disturbance produces only a small response.

Imagine instead that the pencil is stood on its unsharpened end. With care, this can just be done, but the response of the pencil to a push is now qualitatively different. If the pencil is pushed more than a very little, it will begin to topple over, and once this motion has begun it will continue on its own, even if the push is removed, the subsequent motion being driven by the weight of the pencil and needing no other external force to make it go on. The pencil will only come to rest in a configuration completely different from its starting one, lying on the table. This is an example of an unstable response: a small disturbance leads to a large departure of the system from its initial condition. Only if the initial push were very small would this large motion not occur, and to be more precise we might say that the response to large disturbances is unstable, but the response to very small disturbances is stable.

Finally, imagine the pencil balanced on its point. It is now wholly unstable. Any disturbance, no matter how small, will cause the pencil to topple. In reality, a system will not remain in an unstable state of this sort, because small disturbances are inevitably present, and will set off motion away from the unstable state. An unstable system can only be stabilized by active intervention from outside: a circus juggler can balance a walking-stick on his nose, but only by moving his nose so as to counteract the instability of the walking-stick.

Many kinds of buckling are examples of instability. Under a small axial load, the compressed wooden strip is stable. If, for instance, it is acted on by a small lateral load, it responds by bowing sideways a little, but the sideways bowing is limited, depends on the presence of the lateral load, and disappears when the load is removed. At a larger axial load, a very small additional lateral load will produce a large sideways bend, and finally, at a still larger axial load, the strip can depart from its initially straight configuration no matter how small the lateral load. It is then unstable, and the load at which it just becomes unstable is the *critical load* at which buckling is observed, and the strip moves out of its original configuration, which has become unstable. In the same way, when the wide paper cylinder is compressed, its original cylindrical form is stable at low loads, but at larger loads is unstable, so that the paper 'prefers' to move out of its unstable cylindrical form into a stable crinkled form.

The application of these ideas to structural mechanics is best exemplified by the study of a very simple system, one so simple that the equations describing it are always very straightforward, and mathematical detail does not draw our attention away from the mechanics.

10.2. Instability of a simple mechanism

In Figure 10.1(a) a weightless rod OA stands vertically, hinged at its lower end to a rigid base, and held at its upper end by two springs, which are unloaded when the rod is vertical. The other ends of the springs are fixed to guides, which can move up and down so that the ends of the springs are always precisely level with A, so that the change in length of each spring is

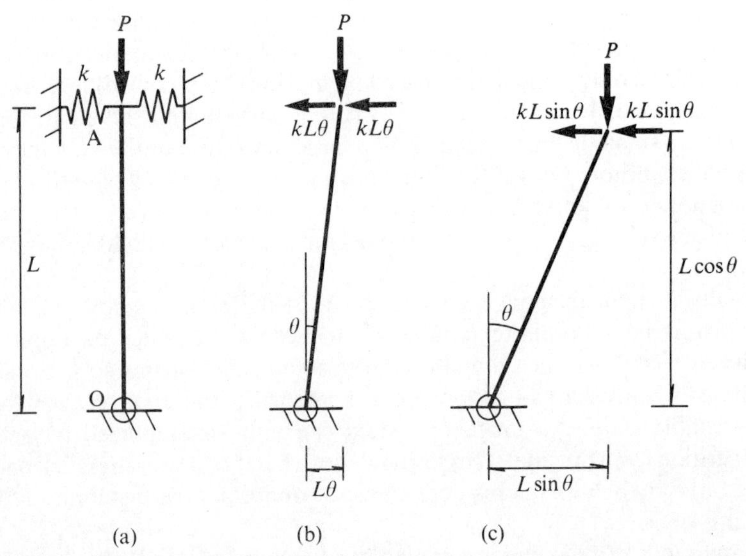

FIG. 10.1. (a) Hinged rod, supported by two springs. (b) Forces induced by a rotation through a small angle θ. (c) Forces induced by a rotation through an angle θ, which is not necessarily small.

equal to the sideways displacement of A. Each spring has a constant stiffness k, so that if its length is changed by e it exerts a restoring force ke opposing the change of length. A vertical load P acts downwards on the rod at A.

The undisturbed rod is certainly in equilibrium: the vertical load P is balanced by a vertical reaction at O, and neither spring exerts a horizontal reaction on the rod. Is this equilibrium stable? We saw that stability has to do with responses to disturbances, and so we examine how the system responds if we move it a little way out of the position where OA is vertical.

Suppose the rod rotated clockwise through an angle θ (Figure 10.1(b)). If θ is small, the horizontal displacement of A is $L\theta$. The left-hand spring is stretched through $L\theta$, and therefore exerts on the rod at A a force $kL\theta$ to the left. The right-hand spring is compressed though $L\theta$, and exerts a force $kL\theta$ to the left, so that together the two springs exert a force $2kL\theta$ to the left. Still assuming that θ is small, take moments about O. The lever arm of the vertical load P is the horizontal displacement of A, which is $L\theta$, and so its moment is $PL\theta$, clockwise. The lever arm of the horizontal restoring spring force $2kL\theta$ is L, and so its moment about O is $2kL^2\theta$. The resultant moment about O is

$$(P-2kL)L\theta$$

clockwise. If θ is zero the moment is zero, and so this position, in which the rod is vertical, is always a possible equilibrium position, though not necessarily a stable one. We now look at what happens when θ is not zero.

If P is zero, the moment is $-2kL^2\theta$, which is counter-clockwise if θ is positive, and therefore tends to restore the rod to its original vertical position, to oppose the disturbance. As we should expect, the equilibrium of the rod is then stable. If P is positive, but less than $2kL$, the moment is counterclockwise, and tends to restore the rod to its upright position, and so the equilibrium is still stable. If P is equal to $2kL$, the resultant moment about O is zero, and it appears that the rod can remain in the disturbed position, whatever the value of θ. If P is greater than $2kL$, the moment is clockwise, and therefore its direction is such as to move the rod even further away from its undisturbed vertical position. Since the response to a small disturbance from the vertical is a movement still further from the vertical, it appears that equilibrium with the rod vertical is unstable if P is greater than $2kL$. If, then, the load P is increased from zero, we can expect the rod to remain vertical when P is less than $2kL$, but to fall over when P reaches the critical load $2kL$, since beyond that load the vertical position is unstable.

Although it certainly simplified the analysis to assume that θ is small, it turns out that we learn more from it if we carry it out exactly, without any assumptions about θ. The exact horizontal displacement of A is $L \sin \theta$ (Figure 10.1(c)), and so each spring exerts a horizontal restoring force $kL \sin \theta$, whose lever arm is $L \cos \theta$, the vertical distance between O and A.

The resultant moment about O is

$$PL \sin \theta - 2kL^2 \sin \theta \cos \theta.$$

If θ is positive, and

$$P > 2kL \cos \theta \tag{10.2.1}$$

the resultant moment is clockwise, and the rod tends to move away from the vertical. If

$$P < 2kL \cos \theta \tag{10.2.2}$$

the resultant moment is counterclockwise, and tends to bring the rod back towards the vertical. The boundary between these conditions is

$$P = 2kL \cos \theta, \tag{10.2.3}$$

a stability boundary between states in which the rod moves away from the vertical and states in which it moves towards the vertical. In Figure 10.2 a

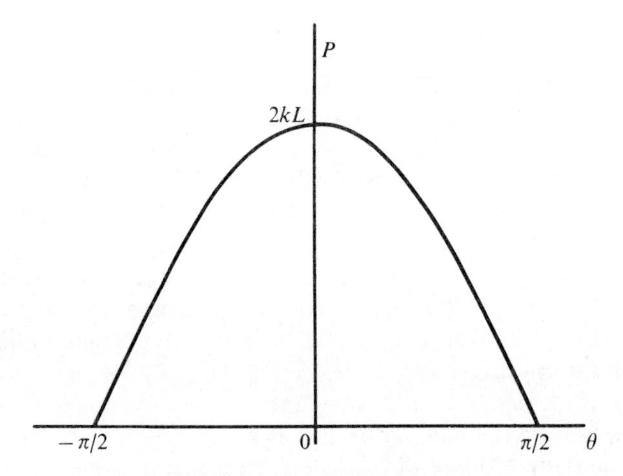

FIG. 10.2. Stability boundary.

point represents a combination of P and θ. If the state of the system is represented by a point inside the stability boundary located by eqn (10.2.3), the rod tends to return to the vertical, but outside the boundary the rod moves away from the vertical. This diagram makes clearer what happens when P reaches $2kL$. If the load is a dead load, such as a weight, which remains unaltered as A moves, the rod is unstable at $2kL$, because any disturbance, no matter how small, will move the point representing the state into the unstable region outside the boundary, and the deflection of the rod will increase.

10.3. Imperfections

Real structures are inevitably imperfect. Columns are not precisely straight, loads are not quite in line, and connections are not freely hinged. These imperfections have little influence on the features of structural behaviour that most of this book has been concerned with. The collapse load of a portal frame, for instance, is scarcely affected even by quite severe misalignment or out-of-straightness of the frame components. However, imperfections do have a significant influence on buckling loads, and their effects can determine the whole behaviour of a structure. This fact has important consequences in civil engineering: an engineer in charge of the fabrication of a box girder must know that if the plates are not flat initially, or if the welds are not in line, or if residual stresses introduced by welding leave the girder twisted slightly, then the load it can carry may be significantly reduced.

Some of the effects of imperfections can be illustrated by the simple mechanism that we studied in Section 10.2. Suppose that when the mechanism is assembled, and the springs are free of tension, the rod is not quite vertical, but at a small angle α to the vertical (Figure 10.3(a)). If a load P is

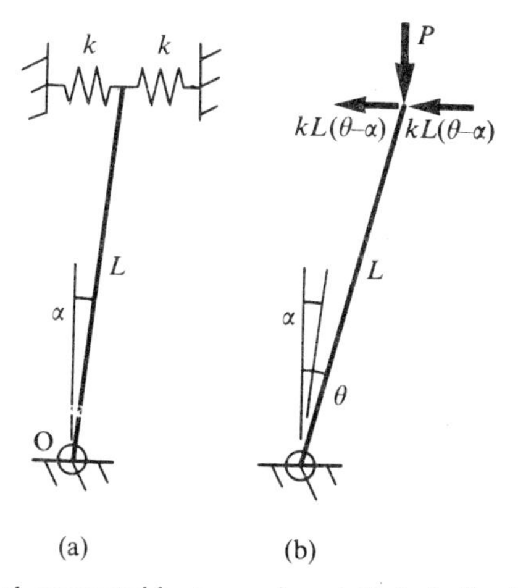

FIG. 10.3. (a) Hinged rod, supported by two springs, initially inclined at an angle α to the vertical. (b) Hinged rod, after rotation to an angle θ to the vertical.

now applied, the rod will rotate clockwise, even under a small load (Figure 10.3(b)). Denote the inclination of the rod to the vertical by θ. Again, to start with, suppose θ and α to be small. The change of length of each spring is $L(\theta-\alpha)$, and so the two springs together exert a restoring force $2kL(\theta-\alpha)$. The resultant moment about O is

$$PL\theta - 2kL^2(\theta-\alpha).$$

If the rod is in equilibrium

$$PL\theta - 2kL^2(\theta - \alpha) = 0. \tag{10.3.1}$$

Rearranging this equation

$$\theta = \frac{\alpha}{1 - \dfrac{P}{2kL}}. \tag{10.3.2}$$

When P is zero, θ is α, as we expect. As P increases, θ increases, at first slowly, but then more and more rapidly. When $P/2kL$ is 0·5, θ is 2α, and so the inclination is twice what it was initially. When $P/2kL$ is 0·8, θ is 5α; when $P/2kL$ is 0·9, θ is 10α. As P approaches $2kL$, which we found to be the critical load for a perfect structure, the inclination of the rod becomes infinitely large.

If we put aside the assumption that θ and α are small, the exact condition for equilibrium, that the moment about O be zero, is

$$PL \sin \theta = 2kL^2(\sin \theta - \sin \alpha) \cos \theta \tag{10.3.3}$$

$$\frac{P}{2kL} = \cos \theta \left(1 - \frac{\sin \alpha}{\sin \theta}\right). \tag{10.3.4}$$

This cannot be solved explicitly to give θ in terms of P, but it is easy to find $P/2kL$ as a function of θ for a particular value of α. In Figure 10.4 this

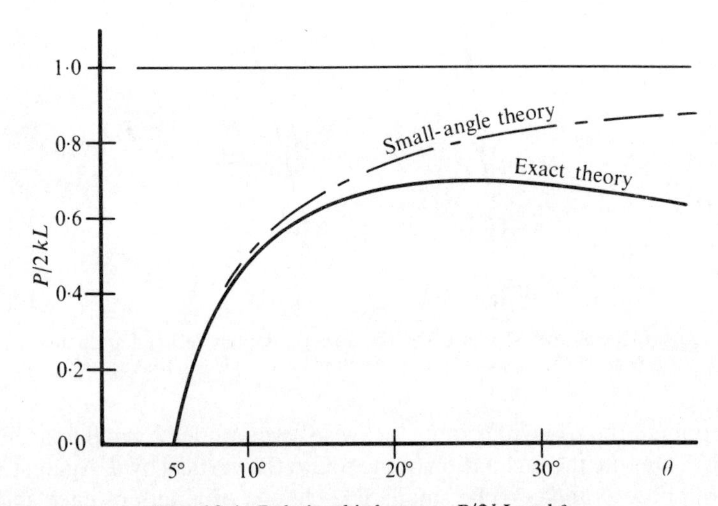

FIG. 10.4. Relationship between $P/2kL$ and θ.

relationship is plotted for α equal to 5°. As the load increases, the inclination increases, at an increasing rate. At a load of 1·44 kL, the inclination is 26·3°.

This is the maximum load the rod can carry, because a larger angle θ corresponds to a smaller equilibrium load. If the load is continuously increased from zero, the rod becomes unstable and falls over when P is $1{\cdot}44\ kL$, which is substantially less than the instability load in the absence of imperfections.

In this instance the solution (10.3.2) given by the small-angle theory is distinctly misleading. It is shown as a chain-dotted line in Figure 10.4. It wrongly predicts that the load can be increased to $2kL$ before deflections increase indefinitely, and indeed the whole form of the relationship is wrong, except at small loads. This is not an isolated case: it quite commonly happens in stability problems that small-deflection assumptions lead to dangerously misleading results. The difficulty is that without these assumptions the mathematical description of the problem is often intractable. This problem is the subject of active research.

10.4. Instability of a column

The mechanism we have considered in the last two sections has been rather an artificial one. We now consider the simplest case of buckling of a structural element, an axially compressed column.

When the rod in Section 10.2 tilted sideways, the lateral deflection of its upper end had to be brought into account when we came to take moments about the hinge, for it was this deflection that gave the load its lever arm about the hinge. That tells us that when we consider the equilibrium of a column that can deflect laterally, we shall need to include similar terms in the equilibrium equations. Consider the equilibrium of an element of a column which is deflected through v from a reference axis x (Figure 10.5). The

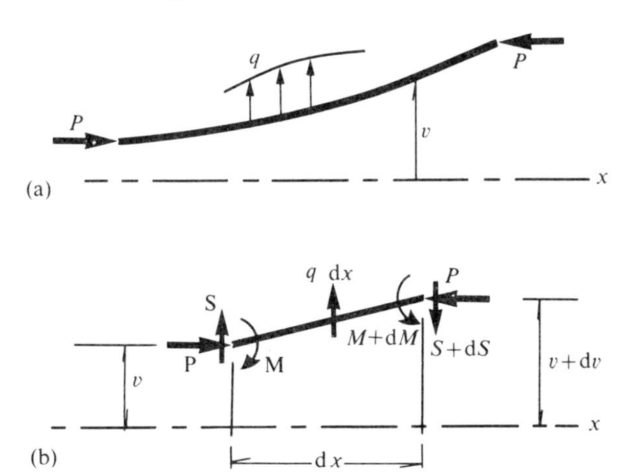

FIG. 10.5. (a) Column under axial load P and lateral load intensity q. (b) Forces and moments on a small element of the column.

deflection v varies from point to point, but in such a way that $\mathrm{d}v/\mathrm{d}x$ is always small. The column carries a distributed lateral load q per unit length, perpendicular to the x-axis, and a compressive force P, whose line of action is parallel to the x-axis. There will as usual be a shear force S and a bending moment M, describing the force and moment exerted by one segment of the beam on another, and they too will vary with x. Figure 10.5(b) shows an element of the compressed beam; it corresponds exactly to Figure 4.8, except that in that case there was no axial force. From the equilibrium of the element,

resolving perpendicular to the x-axis, $S+q\,\mathrm{d}x-(S+\mathrm{d}S)=0,$ (10.4.1)

taking moments, $M+(S+\mathrm{d}S)\,\mathrm{d}x-P\,\mathrm{d}v-(q\,\mathrm{d}x)(\tfrac{1}{2}\,\mathrm{d}x)-(M+\mathrm{d}M)=0.$ (10.4.2)

The term $P\,\mathrm{d}v$ enters the second equation because the forces P at the ends of the element are not in line, and therefore exert a moment.

In the limit, as $\mathrm{d}x$ tends to zero, these equations become

$$\frac{\mathrm{d}S}{\mathrm{d}x}=q \tag{10.4.3}$$

$$\frac{\mathrm{d}M}{\mathrm{d}x}+P\frac{\mathrm{d}v}{\mathrm{d}x}=S. \tag{10.4.4}$$

Equation (10.4.3) is eqn (4.2.3), derived before; eqn (10.4.4) is eqn (4.2.4) with the addition of an extra term accounting for the moment of P. This extra term is negligible in the kinds of problems dealt with in Chapter 7, as you can confirm for yourself by calculation, but is not negligible in stability problems. These two equations express equilibrium.

If the column is linear, bending moment M and change of curvature are related by

$$M=F\kappa. \tag{10.4.5}$$

Suppose the column to be initially straight and parallel to the x-axis. Its change of curvature as it deflects is

$$\kappa=\frac{\mathrm{d}^2v}{\mathrm{d}x^2}\Bigg/\left\{1+\left(\frac{\mathrm{d}v}{\mathrm{d}x}\right)^2\right\}^{\frac{3}{2}}, \tag{10.4.6}$$

and if $\mathrm{d}v/\mathrm{d}x$ is much less than unity in absolute value, which it will be if the deflection is small,

$$\kappa=\frac{\mathrm{d}^2v}{\mathrm{d}x^2} \tag{10.4.7}$$

approximately. We can now combine the curvature–deflection relation (10.4.7), the moment–curvature relation (10.4.5) and the two equilibrium equations. Differentiating eqn (10.4.4) with respect to x (supposing that P is independent of x), and using eqn (10.4.3)

$$\frac{d^2M}{dx^2}+P\frac{d^2v}{dx^2}=q. \tag{10.4.8}$$

Substituting for κ in eqn (10.4.5), and then for M in eqn (10.4.8), and supposing the lateral load q to be zero,

$$F\frac{d^4v}{dx^4}+P\frac{d^2v}{dx^2}=0. \tag{10.4.9}$$

This equation always has the trivial solution

$$v=0 \quad \text{for all } x, \tag{10.4.10}$$

which corresponds to the straight unbuckled form. What we are looking for are conditions under which this form becomes unstable. We saw in the example of the spring-supported hinged rod that at the critical load $2kL$ other equilibrium positions became possible, besides the original one in which the rod was vertical. Following the same strategy here, we look for conditions under which the governing differential equation is satisfied, but the column is not straight. Mathematically, eqn (10.4.9) is obeyed, but there are other solutions in addition to the trivial one.

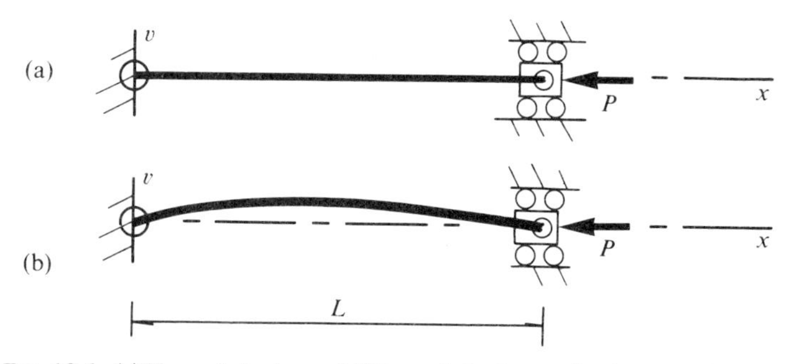

FIG. 10.6. (a) Pin-ended column. (b) Pin-ended column, after lateral deflection.

Consider first of all the column shown in Figure 10.6. Initially the column is straight, pinned at one end to a foundation, pinned at the other end to a support which is free to move in the initial axial direction, but not laterally. Figure 10.6(a) shows the initial configuration, and Figure 10.6(b) a possible deflected configuration under an axial load P. If the column deflects, its deflection must obey eqn (10.4.9), and it must also obey conditions imposed by the constraints at the ends (exactly like the boundary conditions we needed to use in beam deflection problems). At the left-hand end, $x=0$, and

$$v=0, \tag{10.4.11}$$

because the hinge is prevented by the foundation from moving laterally, and

$$M = 0, \tag{10.4.12}$$

because the end is hinged, and therefore the curvature

$$\frac{d^2 v}{dx^2} = 0, \tag{10.4.13}$$

from eqns (10.4.5) and (10.4.7). At the other end, $x = L$,

$$v = 0, \tag{10.4.14}$$

because the support prevents lateral deflection, and

$$\frac{d^2 v}{dx^2} = 0, \tag{10.4.15}$$

because the end is again hinged. We are looking for a condition that allows the column to move out of its straight configuration into a deflected one, and that obeys the differential equation (10.4.9) and the boundary conditions (10.4.11), (10.4.13), (10.4.14), and (10.4.15). The differential equation is of fourth order, linear, with constant coefficients, and its solution can be found in the customary way. If

$$v = c \exp(ax) \tag{10.4.16}$$

is a trial solution, then

$$Fa^4 + Pa^2 = 0, \tag{10.4.17}$$

and the four roots of this equation are

$$a = 0 \text{ (twice)}, +ib \text{ and } -ib, \tag{10.4.18}$$

where

$$b^2 = P/F \tag{10.4.19}$$

and b is positive. It follows that the general solution to eqn (10.4.9) is

$$v = c_1 + c_2 x + c_3 \sin bx + c_4 \cos bx, \tag{10.4.20}$$

where c_1, c_2, c_3, and c_4 are arbitrary constants. If the beam deflects laterally, at least one of these constants must be non-zero. The first and second derivatives are

$$dv/dx = c_2 + bc_3 \cos bx - bc_4 \sin bx \tag{10.4.21}$$

$$d^2v/dx^2 = -b^2 c_3 \sin bx - b^2 c_4 \cos bx. \tag{10.4.22}$$

The four end conditions give

$$0 = c_1 + c_4 \tag{10.4.23}$$

from eqn (10.4.11);

$$0 = -b^2 c_4 \tag{10.4.24}$$

from eqn (10.4.13);

$$0 = c_1 + c_2 L + c_3 \sin bL + c_4 \cos bL \tag{10.4.25}$$

from eqn (10.4.14), and

$$0 = -b^2 c_3 \sin bL - b^2 c_4 \cos bL \tag{10.4.26}$$

from eqn (10.4.15).

Eqn (10.4.24) tells us that c_4 must be zero (since b^2 is non-zero), and then eqn (10.4.23) requires that c_1 be zero as well. Equations (10.4.25) and (10.4.26) then become

$$0 = c_2 L + c_3 \sin bL \tag{10.4.27}$$

and

$$0 = -b^2 c_3 \sin bL. \tag{10.4.28}$$

The second of these requires either that c_3 be zero or that sin bL be zero. If c_3 were zero, eqn (10.4.27) would make c_2 zero, and then all the coefficients would be zero, and we should get back to the trivial undeflected configuration solution, eqn (10.4.10). Accordingly, we must have

$$0 = \sin bL. \tag{10.4.29}$$

The roots of this equation are

$$bL = 0, \pi, 2\pi, 3\pi, \ldots, n\pi, \ldots \tag{10.4.30}$$

and so on. The first root,

$$bL = 0,$$

is of no interest: since b is not zero, it requires that L be zero, that there is no column! The second root is

$$bL = \pi,$$

and so

$$(P/F)^{\frac{1}{2}} L = \pi \tag{10.4.31}$$

and

$$P = \pi^2 F / L^2. \tag{10.4.32}$$

When P has this critical value, the column can deflect sideways. At lower loads, it cannot. Mathematically, at the critical load the differential equation and the boundary conditions are obeyed, and even so there can be a lateral deflection, with c_3 non-zero, though c_1, c_2, and c_4 are zero. This is the critical load for a compressed uniform pin-ended column, often called the Euler load after the great eighteenth-century mathematician who first carried out an analysis of this kind. It corresponds to the critical load $2kL$ for the hinged-rod problem.

Strictly speaking, we have not shown that this is the load at which a column becomes unstable, but only that at this load lateral deflection becomes possible. The magnitude of the lateral deflection is not determined. A more exact analysis, beyond the scope of this book and a good deal more difficult mathematically, does not approximate the curvature κ by $\mathrm{d}^2v/\mathrm{d}x^2$, and shows the true relationship between load and lateral deflection of the column to be the one illustrated in Figure 10.7.

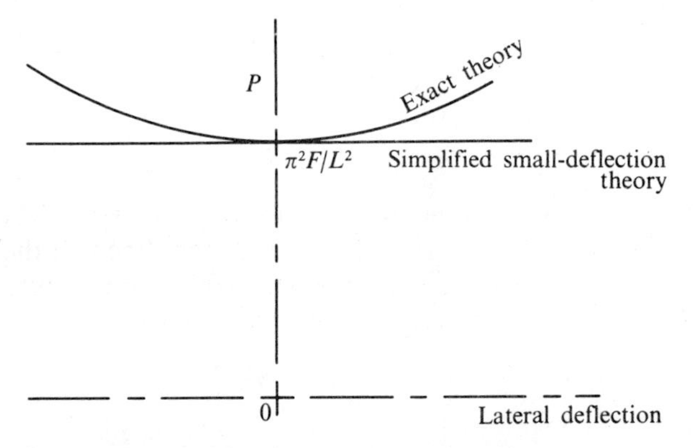

FIG. 10.7. Relationship between axial load and central lateral deflection.

Experimentally, slender columns are found to buckle close to the critical load π^2F/L^2, provided that this load is smaller than the load at which plastic yielding in compression begins to occur. As we expect, increasing the flexural rigidity F of the column increases the buckling load. In the desk-top experiments at the beginning of the chapter we found that reducing the length of a column very much increased its buckling load: that is consistent with the analytical result, which shows that the buckling load is inversely proportional to the square of the length, a result discovered by Musschenbroek in 1729.

Equation (10.4.29) has other roots besides the two we have examined so far, and we ought to pause to see what they correspond to physically. The next root is

$$bL = 2\pi$$

and thence

$$P = 4\pi^2F/L^2;$$

the fourth root is

$$bL = 3\pi$$

and thence

$$P = 9\pi^2 F/L^2;$$

and so on. These higher critical loads $4\pi^2 F/L^2, 9\pi^2 F/L^2, \ldots$ are of no practical importance, because they are larger than the critical load $\pi^2 F/L$. In fact they correspond to different buckling modes in which the column bends into more than one half sine-wave. The first critical load corresponds to the lateral deflection

$$v = c \sin(\pi x/L),$$

the next to

$$v = c \sin(2\pi x/L),$$

the next to

$$v = c \sin(3\pi x/L),$$

and so on.

The derivation above applies to a uniform column with both ends hinged. If the end conditions are different, the buckling load will be different.

Example 10.4.1. A uniform column is shown in Figure 10.8. One end is rigidly built-in to a foundation, and the other end is free. The column carries a load in the direction of its axis. If the loaded free end deflects sideways, the load moves with it, the direction of the load remaining the same. What is the critical value of the compressive load?

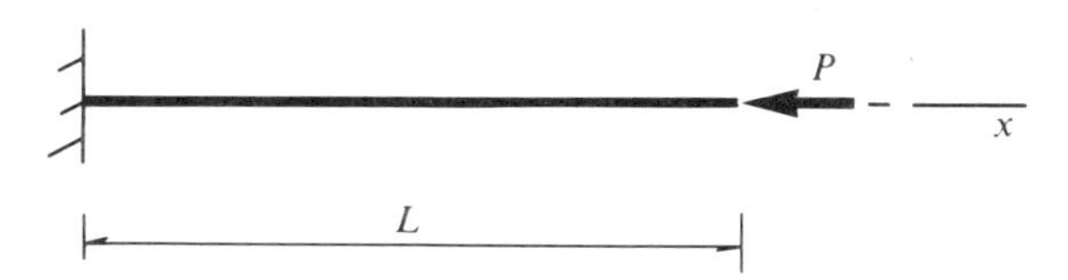

FIG. 10.8. Column, built-in at one end and free at the other end.

The same strategy as before is adopted: we look for conditions under which the column can deflect sideways. Once again, if the column has a linear moment–curvature relation, and a uniform flexural rigidity F, the differential equation is

$$F\frac{\mathrm{d}^4 v}{\mathrm{d}x^4} + P\frac{\mathrm{d}^2 v}{\mathrm{d}x^2} = 0, \tag{10.4.33}$$

but now there are different end conditions. At the left-hand end

$$v = 0 \quad \text{at } x = 0, \tag{10.4.34}$$

because no lateral deflection can occur, and

$$\mathrm{d}v/\mathrm{d}x = 0 \quad \text{at } x = 0, \tag{10.4.35}$$

because the end is built-in. At the other end neither v nor $\mathrm{d}v/\mathrm{d}x$ need be zero, but the bending moment is zero, and so

$$\mathrm{d}^2v/\mathrm{d}x^2 = 0 \quad \text{at } x = L. \tag{10.4.36}$$

That gives us three boundary conditions, and one more is needed. At the left-hand end of the column, the shear force S must be zero, even when the column has deflected, because otherwise the resultant force on the column perpendicular to the x-axis would not be zero. Since

$$S = \frac{\mathrm{d}M}{\mathrm{d}x} + P\frac{\mathrm{d}v}{\mathrm{d}x} \tag{10.4.37}$$

in general, and

$$M = F\frac{\mathrm{d}^2v}{\mathrm{d}x^2}, \tag{10.4.38}$$

it follows that

$$S = F\frac{\mathrm{d}^3v}{\mathrm{d}x^3} + P\frac{\mathrm{d}x}{\mathrm{d}x} \tag{10.4.39}$$

In this instance, at $x = 0$, S is zero, and $\mathrm{d}v/\mathrm{d}x$ is zero by eqn (10.4.35), and so

$$\frac{\mathrm{d}^3v}{\mathrm{d}x^3} = 0 \quad \text{at } x = 0. \tag{10.4.40}$$

The general solution to the governing equation is again

$$v = c_1 + c_2x + c_3 \sin bx + c_4 \cos bx,$$

$$\mathrm{d}v/\mathrm{d}x = c_2 + bc_3 \cos bx - bc_4 \sin bx,$$

$$\mathrm{d}^2v/\mathrm{d}x^2 = -b^2c_3 \sin bx - b^2c_4 \cos bx,$$

$$\mathrm{d}^3v/\mathrm{d}x^3 = -b^3c_3 \cos bx + b^3c_4 \sin bx,$$

and the four boundary conditions, eqns (10.4.34–36) and (10.4.40), give

$$0 = c_1 + c_4, \tag{10.4.41}$$

$$0 = c_2 + bc_3, \tag{10.4.42}$$

$$0 = -b^2c_3 \sin bL - b^2c_4 \cos bL, \tag{10.4.43}$$

$$0 = -b^3c_3. \tag{10.4.44}$$

From the second and fourth of these equations,

$$c_2 = c_3 = 0, \tag{10.4.45}$$

and so the third equation gives

$$c_4 \cos bL = 0. \tag{10.4.46}$$

The constant c_4 cannot be zero, for otherwise c_1 would be zero as well, from eqn (10.4.41), and there would be no deflection. Therefore

$$\cos bL = 0 \tag{10.4.47}$$

$$bL = \frac{\pi}{2}, \frac{3\pi}{2}, \frac{5\pi}{2}, \ldots \tag{10.4.48}$$

The first root gives

$$P = (\pi^2/4)F/L^2, \tag{10.4.49}$$

and that is the critical load in this instance, smaller than the critical load for a pin-ended column because the end is now free to deflect sideways. It is again proportional to the flexural rigidity, and inversely proportional to the square of the length. The buckling mode is

$$v = c_4(\cos(2\pi x/L) - 1). \tag{10.4.50}$$

Example 10.4.2. Suppose the column studied in Example 10.4.1 modified in the way shown in Figure 10.9. The left-hand end is still built-in, but the right-hand end is no longer completely free to move sideways. Instead it is restrained by two springs, each with stiffness k, which exert a restraining force if it deflects sideways, so that the end restraining force is 2k times the end deflection. What is then the critical load?

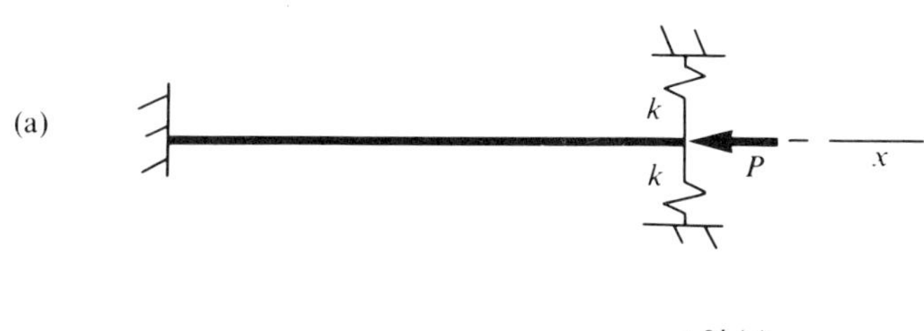

FIG. 10.9. Column, built-in at one end and restrained by linear springs at the other end.

Three of the boundary conditions are unaltered:

$$v = 0 \quad \text{at } x = 0 \tag{10.4.51}$$

$$\mathrm{d}v/\mathrm{d}x = 0 \quad \text{at } x = 0 \tag{10.4.52}$$

$$\mathrm{d}^2v/\mathrm{d}x^2 = 0 \quad \text{at } x = L. \tag{10.4.53}$$

The fourth condition is altered: if the end deflects sideways, the lateral force at that end must be balanced by a shear force at the other end, as in Figure 10.9(b). From equilibrium

$$2k(v)_{x=L} = (S)_{x=0} \tag{10.4.54}$$

$$= F\left\{\left(\frac{\mathrm{d}^3v}{\mathrm{d}x^3}\right)_{x=0} + P\left(\frac{\mathrm{d}v}{\mathrm{d}x}\right)_{x=0}\right\}, \tag{10.4.55}$$

using eqn (10.4.39),

$$= F\left(\frac{\mathrm{d}^3v}{\mathrm{d}x^3}\right)_{x=0} \tag{10.4.56}$$

using the second boundary condition, eqn (10.4.52). This gives us a fourth boundary condition.

The general solution is again eqn (10.4.20), and the four boundary conditions (10.4.51–53, 56) give

$$0 = c_1 + c_4 \tag{10.4.57}$$

$$0 = c_2 + bc_3 \tag{10.4.58}$$

$$0 = -b^2c_3 \sin bL - b^2c_4 \cos bL \tag{10.4.59}$$

$$0 = c_1 + c_2L + c_3(\sin bL + Fb^3/2k) + c_4 \cos bL. \tag{10.4.60}$$

Subtracting eqns (10.4.57) and (10.4.58) multiplied by L from (10.4.60), we have

$$0 = c_3(\sin bL - bL + Fb^3/2k) + c_4(\cos bL - 1) \tag{10.4.61}$$

and dividing eqn (10.4.59) by b^2

$$0 = c_3 \sin bL + c_4 \cos bL. \tag{10.4.62}$$

Each of these last two equations determines c_3/c_4, and this ratio is not indeterminate unless c_3 and c_4 are both zero. Equating the ratios, and rearranging

$$\frac{2kL^3}{F} = \frac{(bL)^3}{bL - \tan bL}. \tag{10.4.63}$$

The column can deflect sideways when this condition is obeyed†. The corresponding value of bL naturally depends on the spring stiffness, which enters the equation through $2kL^3/F$, a dimensionless ratio expressing the relative stiffness of the column and the springs. It is easiest to find the solution graphically. In Figure 10.10 the right-hand side of eqn (10.4.63) is

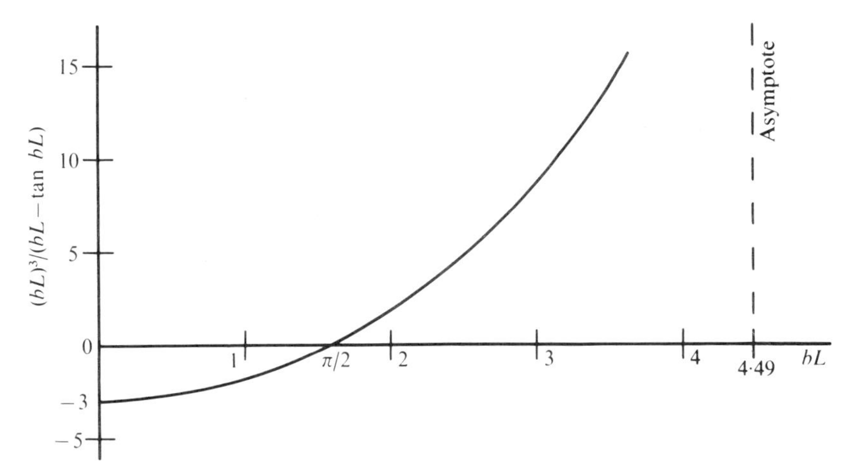

FIG. 10.10. Graph for solution of equation (10.4.63).

plotted as a function of bL. It is negative when bL is less than $\pi/2$, zero when bL is $\pi/2$, and positive when bL is greater than $\pi/2$. The value at which eqn (10.4.63) is obeyed is given by the intersection of the graph of $(bL)^3/(bL - \tan bL)$ and a horizontal line at $2kL^3/F$. If k is zero, as in Example 10.4.1, $2kL^3/F$ is zero, and the intersection is at $bL = \pi/2$, which gives the critical load $(\pi^2/4)F/L^2$ found before. As k increases, $2kL^3/F$ increases, and the intersection moves upwards and to the right, so that the critical value of bL increases, as we should expect. However large k is, the critical load cannot increase beyond the value which makes $(bL)^3/(bL - \tan bL)$ infinite, when

$$bL = \tan bL,$$

which gives

$$bL = 4{\cdot}49,$$

whence

$$P = 20{\cdot}2F/L^2.$$

† An alternative approach, mathematically more sophisticated, is to use the fact that if we consider eqns (10.4.57–60) as four simultaneous equations in the constants c_1, c_2, c_3, and c_4, the condition for the equations to have a solution in which not all the constants are zero is that the determinant of the coefficient matrix be zero, or, equivalently, that the matrix be singular. That gives the same result.

10.5. Initially-bent columns

The hinged-rod problem showed that the behaviour of an imperfect system was qualitatively different from that of a perfect one. Imagine now that the pin-ended column studied in Section 7.4 is initially out-of-straight, so that when it carries no load or bending moment

$$v = v_0(x), \tag{10.5.1}$$

where v_0 is a function measuring the initial deflection from the x-axis, along which the axial load acts. The equilibrium equations (10.4.3), (10.4.4), and (10.4.8) still apply, but the bending moment M is related to the change in curvature from the initial form. The change in curvature κ is

$$\kappa = \frac{\mathrm{d}^2}{\mathrm{d}x^2}(v - v_0) \tag{10.5.2}$$

if the deflection is small. Again assume that the bending moment is linearly related to the change in curvature by a uniform flexural rigidity F, so that

$$M = F\kappa = F\frac{\mathrm{d}^2}{\mathrm{d}x^2}(v - v_0). \tag{10.5.3}$$

Substituting into eqn (10.4.8), the differential equation governing an initially-bent column is

$$F\frac{\mathrm{d}^4}{\mathrm{d}x^4}(v - v_0) + P\frac{\mathrm{d}^2 v}{\mathrm{d}x^2} = q. \tag{10.5.4}$$

Again consider the pin-ended column with no lateral load, so that q is zero, and let it initially be bowed into half a sine wave, so that

$$v_0 = a \sin(\pi x/L), \tag{10.5.5}$$

where a is the central deflection and L is the column length. Then eqn (10.5.4) becomes

$$F\frac{\mathrm{d}^4 v}{\mathrm{d}x^4} + P\frac{\mathrm{d}^2 v}{\mathrm{d}x^2} = a(\pi/L)^4 \sin(\pi x/L). \tag{10.5.6}$$

Its general solution is found by the usual method for linear differential equations, and is

$$v = \frac{a \sin(\pi x/L)}{1 - \dfrac{P}{\pi^2 F/L^2}} + c_1 + c_2 x + c_3 \sin bx + c_4 \cos bx, \tag{10.5.7}$$

where again $b^2 = P/F$ and c_1, c_2, c_3, and c_4 are constants. The first term is the particular integral, and the remaining four together make up the complementary function. The boundary conditions (10.4.11, 13–15) still apply;

substituting this new solution into them, we have

$$0 = c_1 + c_4 \tag{10.5.8}$$

$$0 = -b^2 c_4 \tag{10.5.9}$$

$$0 = c_1 + c_2 L + c_3 \sin bL + c_4 \cos bL \tag{10.5.10}$$

$$0 = -b^2 c_3 \sin bL - b^2 c_4 \cos bL. \tag{10.5.11}$$

We saw earlier that the smallest value of bL at which these equations have a solution other than

$$c_1 = c_2 = c_3 = c_4 = 0$$

was the value corresponding to the critical load $\pi^2 F/L^2$. If the load is smaller than this, all the constants are zero, and

$$v = \frac{a \sin(\pi x/L)}{1 - \dfrac{P}{\pi^2 F/L^2}}, \text{ provided } P < \pi^2 F/L^2. \tag{10.5.12}$$

This is very like eqn (10.3.2). If P is zero, the central deflection is a. As P increases, the denominator decreases, and the central deflection increases, at first slowly and then more rapidly. As P approaches the critical load $\pi^2 F/L^2$, which we previously found independently, the central deflection becomes infinitely large. This, of course, is a result derived in the context of small-deflection theory, and we expect it to fail to describe the true behaviour at very large deflections, just as small-deflection theory proved to be misleading in Section 10.3.

This solution was particularly simple, because the initial deflection v_0 was in the same mode as the buckling mode of an initially-straight column with the same end conditions. Suppose instead that

$$v_0 = \begin{cases} 2ax/L & x < \frac{1}{2}L \\ 2a(1 - x/L) & x > \frac{1}{2}L. \end{cases} \tag{10.5.13}$$

This is an initial deflection in which the column has two straight segments with a kink at the middle, the midpoint being offset a distance a from a line between the end hinges. It is simplest to represent this by a Fourier series, so that

$$v_0 = \sum_{n=1,3,5\ldots} \frac{8a}{\pi^2 n^2} (-1)^{(n-1)/2} \sin(n\pi x/L) \tag{10.5.14}$$

$$= \frac{8a}{\pi^2}\left\{\sin(\pi x/L) - \frac{1}{9}\sin(3\pi x/L) + \frac{1}{25}\sin(5\pi x/L)\ldots\right\}. \tag{10.5.15}$$

The solution to eqn (10.5.4) then turns out to be

$$v = \sum_{n=1,3,5\ldots} \frac{8a}{n^2\pi^2} \frac{(-1)^{(n-1)/2}}{1 - \dfrac{P}{n^2\pi^2 F/L^2}} \sin(n\pi x/L) \tag{10.5.16}$$

$$= \frac{8a}{\pi^2} \left\{ \frac{\sin(\pi x/L)}{1 - \dfrac{P}{\pi^2 F/L^2}} - \frac{\frac{1}{9}\sin(3\pi x/L)}{1 - \dfrac{P}{9\pi^2 F/L^2}} + \frac{\frac{1}{25}\sin(5\pi x/L)}{1 - \dfrac{P}{25\pi^2 F/L^2}} \cdots \right\}. \tag{10.5.17}$$

The first term is like the right-hand side of eqn (10.5.12). Its denominator tends to zero as P approaches the critical load π^2F/L^2; the denominator of the next term tends to zero more slowly, as P approaches the third critical load $9\pi^2F/L^2$; the denominator of the third term tends to zero still more slowly, as P approaches the fifth critical load $25\pi^2F/L^2$, and so on. The consequence is that as P tends to the critical load π^2F/L^2, the factor multiplying the $\sin(\pi x/L)$ term increases from 1 to infinity, but the factor multiplying the next term only increases from 1/9 to 1/8, and so on. The first term, which is much the largest when P is zero, becomes even more dominant as P increases. Once again, the load at which deflections become very large is π^2F/L^2, and the corresponding deflection mode approaches a half sine wave.

10.6. Behaviour of real columns

Imagine an axially loaded pin-ended column made from a steel universal column section. Its cross-section is an 'H' 327 mm broad and 311 mm high, with flanges 25 mm thick. Such a section is typical of heavy columns in steel-framed buildings. As long as it behaves elastically, its flexural rigidity is 26 000 kN m^2†. Table 10.1 shows how the critical load π^2F/L^2 depends on the length of the column. Tests show that the failure loads of long columns are close to π^2F/L^2, and that their buckling mode is as the theory predicts. However, if the column is short, in this case less than about 7 m long, the mean compressive stress level corresponding to the critical load is greater than the yield stress, which for a typical steel column is 300 N/mm^2. A different mode of failure then occurs. If the column is very short, it simply yields plastically in compression, at a so-called squash load equal to the yield stress multiplied by the cross-sectional area. A simple approach to the problem of determining failure loads for columns would be to say that a

† A section of this kind can bend in more than one plane, and different flexural rigidities correspond to different direction planes of bending. Since buckling load is proportional to flexural rigidity, the column will buckle by bending in the plane corresponding to the smallest flexural rigidity, provided that the end conditions allow it to buckle in this preferred direction. In the case of an H section, the smallest flexural rigidity corresponds to bending in a plane perpendicular to the cross-bar of the H. It is this flexural rigidity that should be chosen in buckling calculations. The topic of biaxial bending is discussed in detail in texts on mechanics of solids.

TABLE 10.1

Critical loads for a 327×311 mm universal column

Column length (m)	Critical load (kN)	Mean compressive stress at critical load (N/mm^2)
30	286	14
20	642	32
10	2570	128
8	4020	200
6	(7140)	(355)
4	(10 300)	(512)

Loads and stresses in parentheses indicate that column failure below the Euler critical load can be expected.

column fails when the load reaches the Euler load or the mean stress reaches the yield stress, whichever occurs first. Actually this is oversimplified: complex interaction effects occur when the Euler load is close to the squash load, and the imperfections that are always present in real columns then become particularly important. Another phenomenon is also present. When a length of steel column comes from the rolling mill, its axial tension is of course zero, and so the average stress over a cross-section is zero, but this does not mean that the column is free of internal stress. Internally, it is a redundant structure, like the frameworks we considered in Chapter 6, and self-equilibrating distributions of stress can exist within it. Measurements show that the edges of the flanges of such a column carry large compressive stresses in the axial direction, balanced by axial tensile stresses near the junctions of the flanges and the web†. These residual stresses result from unequal rates of cooling after rolling. When the column carries a compressive load, the accompanying mean compressive stress is superimposed on the locked-in internal stresses which were already present. The result is that the ends of the flanges yield first. This reduces the effective flexural rigidity, and this in turn influences the buckling load.

Although steel columns have been the subject of tens of thousands of tests, understanding of their behaviour has only come slowly, and is still the subject of active research. This is indeed true of instability theory generally. This chapter adopted a simple analytic strategy, that of looking for loads at which the column can move out of its initial configuration and still remain in equilibrium. In many structural problems, this attractively simple approach is misleading, and historically it in fact held back the understanding of unstable structural behaviour. Important steps forward were made by Shanley (1947) and Kármán and Tsien (1939), who set out to find alternative

† In a steel cross-section in the form of an 'H', the *flanges* are the verticals of the H, and the *web* is the bar.

ways of looking at the problem. Their work goes beyond the scope of this book, but you will find their original papers readable and lively (and largely non-mathematical).

10.7. Problems

1. Figure 10.11 illustrates three structures, each made of one or more rigid rods held by linear springs. Each carries a vertical load. Assuming deflections to be small, determine the loads at which the structures become unstable.

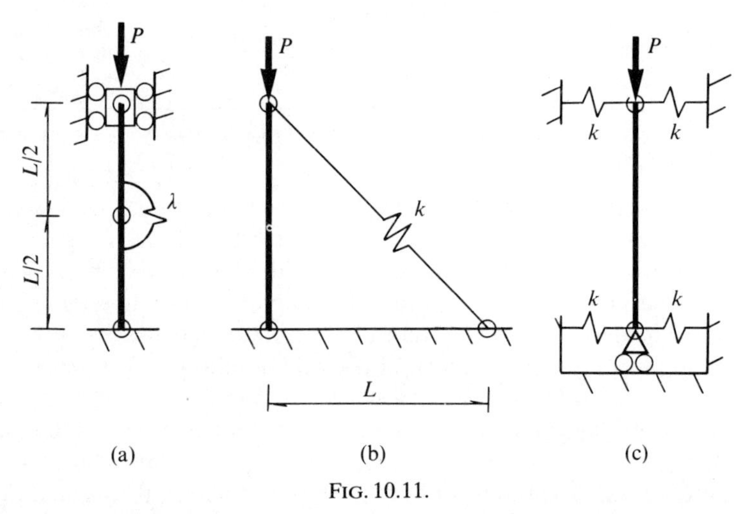

FIG. 10.11.

Structure (a) in the Figure is made of two rigid rods, connected by a hinge which exerts a restoring torque $\lambda\theta$ when the relative rotation between the rods is θ, measured from the position in which they are in line. In structures (b) and (c), each of the springs has a stiffness k, and carries no force in the undeflected position shown.

2. Reinvestigate the behaviour of structure (b) in Figure 10.11, dropping the assumption that the rotation of the rod from the vertical is small. What happens if the rod is initially inclined α out of the vertical?

3. A uniform column of length L and flexural rigidity F is held in such a way that its ends can neither rotate nor deflect sideways. It carries an axial compressive load. Show that the critical value of this load is $4\pi^2 F/L^2$.

4. A uniform cantilever beam of length L and flexural rigidity F has one end clamped. The free end is subjected to a transverse load W, and to a compressive load P whose line of action is always parallel to the original axis of the beam. Derive an expression for the transverse deflection of the free end of the cantilever, and determine the value of P for which this becomes very large.

5. A pin-ended column of length L is non-uniform in section. The middle third has a uniform flexural rigidity F, but the end thirds are effectively rigid.
   (a) How would you expect the critical load for this column to relate to the critical load for a column of length L and uniform flexural rigidity F throughout? Will it be much larger, slightly larger, the same, slightly smaller, or much smaller?
   (b) Find the critical load.

6. In Figure 10.12 a uniform linear column is compressed by forces which are offset by a distance e from the column axis, e being small by comparison with the column length. Find the relation between the load and the central deflection of the column. What happens if the forces are tensile?

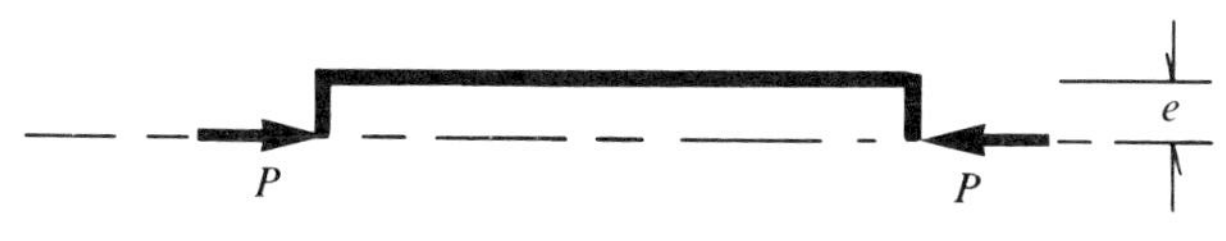

FIG. 10.12.

7. A simply supported uniform beam of length L carries a sinusoidally distributed load of intensity

$$w = w_0 \sin(\pi x/L),$$

where x is the distance from one end and w_0 is constant. It also carries a compressive axial force P. Determine the relationship between the axial force and the central deflection of the beam. Go on to consider the behaviour of an axially compressed beam under a uniformly distributed lateral load.

Appendix: the lower-bound theorem

Chapter 6 the theorem was stated but not proved. The following proof applies to a plane frame which collapses by the development of bending deformation. The proof can easily be generalized.

Before we go on to the main proof, a subsidiary result is needed. Collapse was defined as a condition in which the frame can continue to deform even though the loads are constant. The first thing to prove is that as the frame collapses, the bending moments within it everywhere remain constant. The simple examples worked out in Chapter 6 show this behaviour, and it will now be shown that it is true whenever a frame collapses.

Imagine there to take place an increment of the collapse deformation, from configuration 1 at the start of the increment to configuration 2 at the end. During this deformation, there are changes $\mathrm{d}u_i$ and $\mathrm{d}v_i$ in the displacements of points in the frame, and there are changes in curvature $\mathrm{d}\kappa$ within its elements; the changes in displacements are compatible with the changes in curvature. At the same time, there are no changes in the external loads, because the structure is collapsing, but there may be changes $\mathrm{d}M$ in the bending moments at different parts of the frame; the changes in bending moment are in equilibrium with the zero changes in external loads. Now apply virtual work, choosing as the equilibrium system the changes in the bending moment and the zero changes in external loads, and choosing as the compatible system the changes in displacements and the changes in curvature. Then, as usual

$$\int_{\substack{\text{whole}\\\text{structure}}} (\text{loads})(\text{displacements}) = \int_{\substack{\text{whole}\\\text{structure}}} (\text{moment})(\text{curvature})\,\mathrm{d}s \tag{A.1}$$

in general, and so in this instance

$$0 = \int_{\substack{\text{whole}\\\text{structure}}} (\mathrm{d}M\,\mathrm{d}\kappa)\,\mathrm{d}s, \tag{A.2}$$

since the integrand on the left of eqn (A.1) is zero, because the changes of load are zero.

Figure A.1 shows the relationship between the bending moment and the curvature of a beam made from an elastic perfectly-plastic material. Suppose that during the collapse increment the point representing the state at a

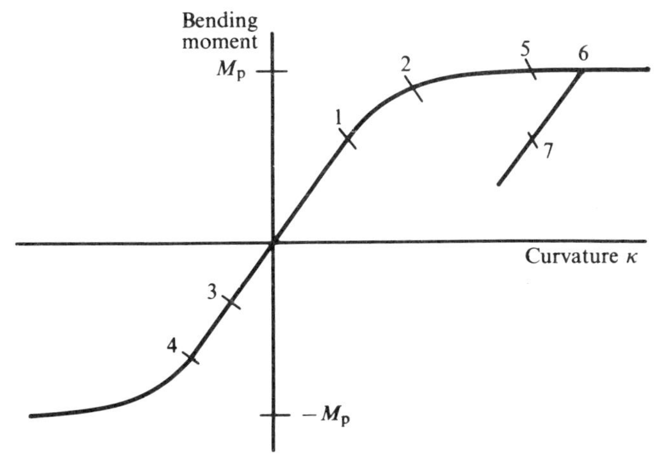

FIG. A.1.

particular section of the beam moves from 1 to 2. Then

$$dM > 0,$$

$$d\kappa > 0,$$

and so

$$dM\, d\kappa > 0.$$

If the beam had bent the other way, so that the point moved from 3 to 4,

$$dM < 0,$$

$$d\kappa < 0,$$

and so

$$dM\, d\kappa > 0.$$

If, however, the moment M is the full plastic moment M_p, and the point representing the state moved from 5 to 6, then (and only then)

$$dM = 0,$$

even though

$$d\kappa > 0,$$

and

$$dM\, d\kappa = 0.$$

Finally, if the beam had previously been bent plastically in one direction, so that its state is represented by point 6, but is now bent the other way, its state follows an unloading line, from 6 to 7, and once again $dM \, d\kappa$ is positive.

It follows that $dM \, d\kappa$ is always positive, unless dM is zero, in which case $dM \, d\kappa$ is zero; $dM \, d\kappa$ can never be negative. Since the integral in eqn (A.2) is zero, and the integrand is never negative, the integrand must everywhere be zero: there is no way in which negative values of $dM \, d\kappa$ in one part of the structure can balance positive values elsewhere. Accordingly

$$dM = 0$$

throughout the structure, so that the bending moment is constant, as well as the external loads, and $d\kappa$ is zero except where M is equal to the local full plastic moment.

We can now go on to the main theorem. It is proved by the method called *reductio ad absurdum.* We assume that the theorem is false, and show that the assumption that the theorem is false inevitably leads to a contradiction: it follows that the theorem must be true.

The loads acting on the frame are U_a and V_a at point A, U_b and V_b at point B, and so on. We can find a distribution of bending moment M^*, in equilibrium with these loads, such that

$$-M_p < M < M_p \tag{A.3}$$

everywhere in the frame. If the lower-bound theorem is true, the frame cannot collapse under the loads.

Imagine that the theorem is false. The frame does collapse, and in an increment of the collapse mechanism point A has displacements u_a and v_a, point B has displacements u_b and v_b, and so on. The corresponding change in curvature is $d\kappa$. The actual bending moment during the collapse is M^{**}, and is in equilibrium with the loads. Apply the theorem of virtual work, with the loads U_a, V_a, . . . and the collapse bending moment M^{**} as the equilibrium system, and the displacements u_a, v_a, . . . and the curvature increments $d\kappa$ as the compatible system. Then

$$\sum_{\substack{\text{whole}\\\text{structure}}} (U_a u_a + V_a v_a + U_b u_b + \cdots) = \int_{\substack{\text{whole}\\\text{structure}}} M^{**} \, d\kappa \, ds. \tag{A.4}$$

Apply virtual work a second time. The loads U_a, V_a, . . . can also be in equilibrium with moments M^*, and so that can be the equilibrium system. Choose the same compatible system as before, displacements u_a, v_a, . . . compatible with curvature increments $d\kappa$. Then

$$\sum_{\substack{\text{whole}\\\text{structure}}} (U_a u_a + V_a v_a + U_b u_b + \cdots) = \int_{\substack{\text{whole}\\\text{structure}}} M^{*} \, d\kappa \, ds. \tag{A.5}$$

Subtracting eqn (A.5) from eqn (A.4)

$$0 = \int_{\substack{\text{whole}\\\text{structure}}} (M^{**} - M^{*})\,\mathrm{d}\kappa\,\mathrm{d}s. \qquad \text{(A.6)}$$

It has been shown that $\mathrm{d}\kappa$ must be zero unless the actual bending moment M^{**} is equal to the full plastic moment. If

$$M^{**} = M_{\mathrm{p}}$$

then

$$M^{**} - M^{*} > 0$$

by inequality (A.3). In addition, $\mathrm{d}\kappa$ must be positive (corresponding to an increment like 5 to 6 in Figure A.1) or zero, and so

$$(M^{**} - M^{*})\,\mathrm{d}\kappa > 0$$

unless $\mathrm{d}\kappa$ is zero. If, alternatively,

$$M^{**} = -M_{\mathrm{p}}$$

then

$$M^{**} - M^{*} < 0$$

and $\mathrm{d}\kappa$ must be negative or zero, and so

$$(M^{**} - M^{*})\,\mathrm{d}\kappa > 0$$

unless $\mathrm{d}\kappa$ is zero.

Now return to equation (A.6). The integral is zero, and we have just shown that the integrand can never be negative. The only possibility is that the integrand is zero everywhere. That implies that $\mathrm{d}\kappa$ must be zero everywhere. If that is so, the structure cannot collapse, because its curvature cannot change. The assumption that the theorem is false, that the structure can collapse even though the conditions of the theorem are obeyed, has led us to the contradictory conclusion that the structure cannot collapse. If the assumption that the theorem is false leads to the contradiction, the theorem must be true.

Answers to problems

CHAPTER 2

1. 12 kN, 48 kN.
2. 36 kN, 24 kN; 8·75 kN.
3. −70·9, 45·0 kN; the sum of the horizontal reaction components is 96·6 kN, but they are not known individually.
4. H must be between 23 and 86 kN.
5. (Figure B.1).

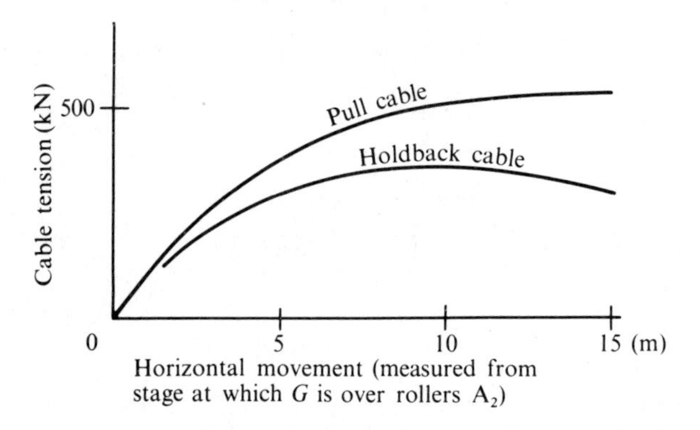

FIG. B.1.

6. −75 kN, −125 kN, 200 kN.
7. at A, 2·59 kN upward; at B, 17·32 kN upward; at C, 4·91 kN downward.

CHAPTER 3

1.

Bar	AB	BC	CD	DE
tension induced by loading (a)	$-3P/4\surd 3$	$-P/4\surd 3$	$P/2\surd 3$	$P/2\surd 3$
tension induced by loading (b)	$3Q/4$	$Q/4$	$-Q/2$	$-Q/2$

Bar	EA	BE	BD
tension induced by loading (a)	$3P/2\surd 3$	$P/2\surd 3$	$-P/2\surd 3$
tension induced by loading (b)	$Q/2$	$-Q/2$	$Q/2$

2. bars BD, EG, HJ, tension +83·3 kN; BF, EI, HK, −83·3; EF, HI, +50; ED, HG, −50; DG, +66·7; GJ, +133·3; EI, −66·7; IK, −133·3; AD, AB, CF, 0; BC, +100.
3. vertical $-(n-i)/n$; diagonal $+(n-i)(a^2+b^2)^{1/2}/nb$
   upper chord $-(n-i)ja/nb$; lower chord $(n-i)(j-1)a/nb$.
4. T_{dc}, 20·62 kN; T_{bd}, −15; T_{de}, −10; T_{be}, 18·03; T_{ab}, −10; T_{bc}, 0.

CHAPTER 4

1. (a) 25/3 kN, 275/3 kN (b) −5 kN m
   (c) $-10x+(25/3)[x-1]^0+(275/3)[x-4]^0$,
   $-5x^2+(25/3)[x-1]+(275/3)[x-4]$, where [ ] are Macauley brackets.
2. (a) $qL/6$, $qL/3$ (b) $S=qL/6-qx^2/2L$, $M=qLx/6-qx^3/6L$
   (c) at $x=L/\sqrt{3}$, $M=qL^2/9\sqrt{3}$.
3. (a) at left-hand abutment, vertical $3qL/8$, horizontal $qL^2/16h$
   at right-hand abutment, vertical $qL/8$, horizontal $qL^2/16h$.
   (b) $+qL^2/64$ (c) $-qL^2/64$.
4. (a) vertical 120 kN, horizontal 40 kN, moment 80 kN m counter-clockwise.
   (b) −40 kN, −25 kN m, +20 kN.
   (c) −80 kN, 20–40y kN m, 40 kN.
   (d) −40 kN, $-200+120x-20x^2$ kN m, $120-40x$ kN.
5. $-16-2x+42[x-4]^0$, $-16x-x^2+42[x-4]$ kN m, 8 kN m.

CHAPTER 5

1. (Figure B.2).

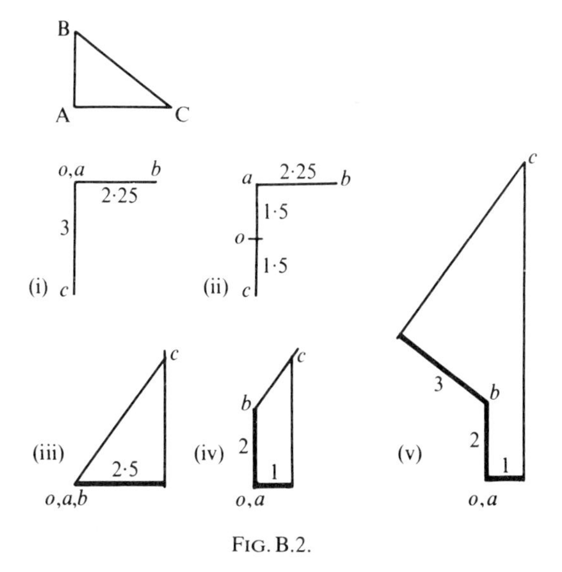

FIG. B.2.

2. (i) 2·31 mm (ii) 1·24 mm.
3. (i) 9 mm (ii) 10·8 mm (iii) 11·67 mm.

4. (i) 120 kN at A, 280 kN at B.
   (ii) $S = 120 - 100[x-5] + 100[x-9]$
   $M = 120x - 50[x-5]^2 + 50[x-9]^2$.
   (iii) 672 kN m, at $x = 6{\cdot}2$ m.
   (iv) 62·5 mm, at $x = 5{\cdot}38$ m.
5. 240 mm.
6. If the bending moment at each end of the beam is denoted $-C$,

$$M = -C + wLx/2 - wx^2/2$$

and the use of the condition that the slope dv/dx is zero at both ends gives $C = wL^2/12$; the central deflection is $wL^4/384F$.

CHAPTER 6

1. (i) 2·31 mm (ii) 1·24 mm.
2. $5WK$, $4WK/\sqrt{3}$.
3. −0·19 mm, −1·31 mm, −4·59 mm.
4. −1·155 kN, −1·155 kN, −2·310 kN.
5. −1·73 kN.
6. $3{\cdot}6\ WL/aE$.

CHAPTER 7

1. (i) $PL^3/3F$ (ii) $5wL^4/384F$.
2. (i) 9 mm (ii) 10·8 mm (iii) 11·67 mm.
3. $(\pi/2)PR^3/F$, $2PR^2/F$; $2CR^2/F$, $\pi CR/F$.
4. (i) $$\kappa = \begin{cases} \{\tfrac{1}{2}wx(L-x) + Rx\}/F & x < L, \\ R(2L-x)/F & x > L, \end{cases}$$

   (ii) $$M = \begin{cases} x & x < L, \\ 2L - x & x > L. \end{cases}$$

   (iii) $-wL/16$, $7wL/16$, $5wL/8$.
   (iv) $7wL^4/768F$.
5. upper right-hand corner, $11pL^2/40$.
6. $$Q\int_0^{L/2} \frac{1}{F(x)}\left(1-\left(\frac{2x}{L}\right)^2\right)^2\left(1+16\left(\frac{h}{L}\right)^2\left(\frac{2x}{L}\right)^2\right)^{\frac{1}{2}} dx$$

$$= \frac{PL}{4h}\int_0^{L/2} \frac{1}{F(x)}\left(1-\left(\frac{2x}{L}\right)^2\right)\left(1-\frac{2x}{L}\right)\left(1+16\left(\frac{h}{L}\right)^2\left(\frac{2x}{L}\right)^2\right)^{\frac{1}{2}} dx.$$

If the assumption that h is small compared to L is not made, the integrals can be evaluated either numerically, or by writing them in terms of elliptic integrals, or by expanding the integrands as power series in h/L and integrating them term-by-term.

7. radial bars $-\alpha\theta AE/12$, circumferential bars $\alpha\theta AE/12$.
8. 3·7 kN.
9. $3\alpha F\theta/L^2$, $2\alpha F\theta/L$.
10. (a) 1 (b) 3 (c) 3 (d) 3 (e) 18 (f) 3.

CHAPTER 8

1. 12 $F\Delta/L^3$, where L is the length of the beam and F its flexural rigidity.
2. reactions $\pm Ph/L$; moment in stanchion $Py/2$ at y above foot, moment in beam $Ph/2 - Phx/L$ at x to the right of the upper left-hand corner.
3. reactions $\pm Ph^2/2L(h+\frac{1}{6}L)$; moment $-(Ph/4)(h+\frac{1}{3}L)/(h+\frac{1}{6}L)$ at left stanchion foot, linear variation along stanchion to $Ph^2/4(h+\frac{1}{6}L)$ at upper end, linear variation along beam to $-Ph^2/4(h+\frac{1}{6}L)$ at other end.
4. (i) $(1/4)C_1L/F$ (ii) $(7/24)C_1L/F$ (iii) $(1/2\sqrt{3})C_1L/F$.
5. 13/32.
7. 6·2 kN, 4·5 kN, 39 kN m.

CHAPTER 9

1. $4M_p/L$.
2. $M_p/Ly(1-y)$.
3. $M_p(1+y)/Ly(1-y)$.
4. $\max\left\{WzL, (WL/2)\dfrac{1-2z}{3-4z}\right\}$; $z=\frac{1}{2}-\frac{1}{4}\sqrt{2}$; $M_P = wL(\frac{1}{2}-\frac{1}{4}\sqrt{2})$.
5. 1·62.
6. $5M_0/L$.
7. 4·95 M_0/L^2; 4·45 M_0/L, 8·92 M_0/L, 1·48 M_0/L.
8. 67 kN; 0·5 m/s.
9. (i) −105 kN (ii) 79·8 kN in BC, −92 kN in CD (iii) 79·8 kN at A, 92 kN at E.
10. Collapse occurs when $Uh/M_p = 2$ or $VL/M_p + 2(Uh/M_p) = 8$.
11. 1·52.
12. The simple 'sidesway' mechanism corresponds to collapse when

$$Uh/M_p = 2.$$

A mechanism with hinges at A, C, D, and E corresponds to collapse when

$$Uh/M_p + \tfrac{1}{4}(VL/M_p) = 4.$$

A general mechanism, with a beam hinge at yL from the left-hand end, and hinges at A, D, and E, gives

$$Uh/M_p + \tfrac{1}{2}y(VL/M_p) = 2/(1-y).$$

Different values of y give a family of lines on the interaction diagram, and these lines have a parabolic envelope

$$Uh/M_p + \tfrac{1}{2}(VL/M_p) = 2(VL/M_p)^{\frac{1}{2}},$$

which describes the collapse condition for $Uh/VL < \frac{1}{2}$.

13. $4(\sqrt{2}+1)M_0/R$.

CHAPTER 10

1. (a) $4\lambda/L$ (b) $\frac{1}{2}kL$ (c) kL.
4. $W(\tan\alpha L - \alpha L)/\alpha P$, where $\alpha^2 = P/F$; $\pi^2F/4L^2$.
5. 15·4 F/L^2.
6. central deflection is $e \sec\dfrac{L}{2}\sqrt{\dfrac{P}{F}}$.
7. $w_0L^4/\pi^4F(1-PL^2/\pi^2F)$.

References

BAKER, J. F., HORNE, M. R., and HEYMANN, J. (1956). *The steel skeleton, Vol. II.* Cambridge University Press.
CALLADINE, C. R. (1969). *Engineering plasticity.* Pergamon Press, Oxford.
CRANDALL, S. H. and DAHL, N. C. (1959). *Introduction to the mechanics of solids.* McGraw-Hill, New York.
DRUCKER, D. C. (1967). *Introduction to mechanics of deformable solids.* McGraw-Hill, New York.
HEYMAN, J. (1964). *Beams and framed structures.* Pergamon Press, Oxford.
HEYMAN, J. (1971). *Plastic design of frames. Vol. II. Applications.* Cambridge University Press.
MACAULEY, W. H. (1919). *Mess. Math.* **48**, 129.
PARKES, E. W. (1965). *Braced frameworks.* Pergamon Press, Oxford.
SHANLEY, F. R. (1947). *J. Aero. Sci.* **14**, 261.
VON KARMAN, T. and TSIEN, H. S. (1939). *J. Aero. Sci.* **7**, 43.

Index